Accelerating Energy Innovation

**A National Bureau of
Economic Research
Conference Report**

Accelerating Energy Innovation
Insights from Multiple Sectors

Edited by **Rebecca M. Henderson and Richard G. Newell**

The University of Chicago Press

Chicago and London

REBECCA M. HENDERSON is the Senator John Heinz Professor of Environmental Management at Harvard Business School, and a research associate of the National Bureau of Economic Research. RICHARD G. NEWELL is the Gendell Associate Professor of Energy and Environmental Economics at the Nicholas School of the Environment, Duke University, and university fellow of Resources for the Future. He currently serves as administrator of the U.S. Energy Information Administration, and was a research associate of the National Bureau of Economic Research prior to the start of his government service.

The University of Chicago Press, Chicago 60637
The University of Chicago Press, Ltd., London
© 2011 by the National Bureau of Economic Research
All rights reserved. Published 2011.
Printed in the United States of America

20 19 18 17 16 15 14 13 12 11   1 2 3 4 5
ISBN-13: 978-0-226-32683-2 (cloth)
ISBN-10: 0-226-32683-7 (cloth)

Library of Congress Cataloging-in-Publication Data

Accelerating energy innovation : insights from multiple sectors / edited by Rebecca M. Henderson and Richard G. Newell.
        p. cm.—(National Bureau of Economic Research conference report)
    ISBN-13: 978-0-226-32683-2 (cloth : alk. paper)
    ISBN-10: 0-226-32683-7 (cloth : alk. paper) 1. Energy industries—Technological innovations. 2. Energy industries—Technological innovations—Economic aspects. 3. Energy policy. 4. Technological innovations—Government policy. I. Henderson, Rebecca. II. Newell, Richard G., 1965– III. Series: National Bureau of Economic Research conference report.
HD9502.A2A324 2011
333.79—dc22

                                                    2010048817

Relation of the Directors to the
Work and Publications of the
National Bureau of Economic Research

1. The object of the NBER is to ascertain and present to the economics profession, and to the public more generally, important economic facts and their interpretation in a scientific manner without policy recommendations. The Board of Directors is charged with the responsibility of ensuring that the work of the NBER is carried on in strict conformity with this object.

2. The President shall establish an internal review process to ensure that book manuscripts proposed for publication DO NOT contain policy recommendations. This shall apply both to the proceedings of conferences and to manuscripts by a single author or by one or more co-authors but shall not apply to authors of comments at NBER conferences who are not NBER affiliates.

3. No book manuscript reporting research shall be published by the NBER until the President has sent to each member of the Board a notice that a manuscript is recommended for publication and that in the President's opinion it is suitable for publication in accordance with the above principles of the NBER. Such notification will include a table of contents and an abstract or summary of the manuscript's content, a list of contributors if applicable, and a response form for use by Directors who desire a copy of the manuscript for review. Each manuscript shall contain a summary drawing attention to the nature and treatment of the problem studied and the main conclusions reached.

4. No volume shall be published until forty-five days have elapsed from the above notification of intention to publish it. During this period a copy shall be sent to any Director requesting it, and if any Director objects to publication on the grounds that the manuscript contains policy recommendations, the objection will be presented to the author(s) or editor(s). In case of dispute, all members of the Board shall be notified, and the President shall appoint an ad hoc committee of the Board to decide the matter; thirty days additional shall be granted for this purpose.

5. The President shall present annually to the Board a report describing the internal manuscript review process, any objections made by Directors before publication or by anyone after publication, any disputes about such matters, and how they were handled.

6. Publications of the NBER issued for informational purposes concerning the work of the Bureau, or issued to inform the public of the activities at the Bureau, including but not limited to the NBER Digest and Reporter, shall be consistent with the object stated in paragraph 1. They shall contain a specific disclaimer noting that they have not passed through the review procedures required in this resolution. The Executive Committee of the Board is charged with the review of all such publications from time to time.

7. NBER working papers and manuscripts distributed on the Bureau's web site are not deemed to be publications for the purpose of this resolution, but they shall be consistent with the object stated in paragraph 1. Working papers shall contain a specific disclaimer noting that they have not passed through the review procedures required in this resolution. The NBER's web site shall contain a similar disclaimer. The President shall establish an internal review process to ensure that the working papers and the web site do not contain policy recommendations, and shall report annually to the Board on this process and any concerns raised in connection with it.

8. Unless otherwise determined by the Board or exempted by the terms of paragraphs 6 and 7, a copy of this resolution shall be printed in each NBER publication as described in paragraph 2 above.

Contents

Acknowledgments

The authors gratefully acknowledge funding from the Industrial Performance Center at the Massachusetts Institute of Technology under a grant from the Doris Duke Charitable Foundation, and from an anonymous donor.

Introduction and Summary

Rebecca M. Henderson and Richard G. Newell

Reorienting current energy systems toward a far greater reliance on technologies with low or no carbon dioxide emissions is an immense challenge. Fossil fuels such as oil, coal, and natural gas together satisfy 81 percent of global energy demand and generate 69 percent of global anthropogenic greenhouse gas (GHG) emissions (International Energy Agency [IEA] 2009). Moreover, worldwide demand for energy is expected to increase by about 40 percent over the next twenty years, with most of this increase occurring in non-Organization for Economic Cooperation and Development (OECD) countries (IEA 2009, EIA 2010).

Meeting this demand without significantly increasing global emissions using currently available technology would be costly. For example, the IEA projects that about $26 trillion of investment in energy-supply infrastructure will be needed over the 2008 to 2030 period simply to meet projected increases in energy demand (IEA 2009).[1] Modeling scenarios of cost-effective global climate mitigation policy suggest that, for atmospheric stabilization targets in the range of 450 to 550 parts per million (ppm) CO_2, the cost of GHG mitigation through 2050 *without significant innovation in the underlying technologies* would require additional trillions or tens of tril-

Rebecca M. Henderson is the Senator John Heinz Professor of Environmental Management at Harvard Business School, and a research associate of the National Bureau of Economic Research. Richard G. Newell is the Gendell Associate Professor of Energy and Environmental Economics at Duke University's Nicholas School of the Environment, and university fellow of Resources for the Future. He currently serves as administrator of the U.S. Energy Information Administration (EIA), and was a research associate of the NBER prior to the start of his government service.

1. This figure does not include expenditures on energy demand-side technologies (e.g., transportation, appliances, and equipment), investment demand for which will measure in the trillions of dollars each year.

lions of dollars (Newell 2008). Longer-term total costs through 2100 could be approximately double this amount.

Plausible developments in energy efficiency, bioenergy, wind, solar, nuclear and low-emitting fossil fuel technologies could greatly reduce these costs. For example, Edmonds et al. (2007) suggest that significant innovation could reduce the present-value cost of achieving CO_2 stabilization at 550 ppm by more than $20 trillion. Other studies have found that the cumulative costs of achieving any given stabilization target are reduced by 50 percent or more under advanced technology scenarios (see, e.g., Manne and Richels 1992; Clarke et al. 2006). Accelerating innovation and technology adoption in energy is, thus, crucial to meeting greenhouse gas mitigation goals.

Given this urgency, it is not surprising that a flood of recent books and articles have explored how the U.S. energy innovation system could be improved (see, for example, work by Anadon and Holdren 2009; Gallagher, Holdren, and Sagar 2006; Anadon, Gallagher, and Bunn 2009; Grübler, Nakićenović, and Victor 1999; Lester 2009; Narayanamurti, Anadon, and Sagar 2009a, b; Newell 2008; Nemet and Kammen 2007; Ogden, Podesta, and Deutch 2008; and Weiss and Bonvillian 2009). By and large, this is tightly argued and persuasive work that is a critically important starting point for anyone interested in effective energy policy. This book explores the same questions using a complementary approach. Instead of focusing on the history of the energy industry to draw lessons for the future of energy innovation, it explores the history of innovation in several industries that have already seen extraordinary rates of technological progress: agriculture, chemicals, semiconductors, computers, the Internet, and biopharmaceuticals. Each of the chapters that follow explores the complex role that public policy and private markets have played in triggering rapid innovation in the industry and in sustaining it once in motion. Each industry differs from the energy sector in important ways, but we nonetheless believe that this approach provides a useful complement to the existing literature.

In the first place, the history of each industry reminds us that relatively rapid, transformational innovation can occur. Innovation in agriculture was a critical factor in reducing the manpower needed to grow the nation's crops from 49 percent of the U.S. workforce in 1880 to less than 2 percent in 2000. The chemical industry created entirely new materials and fuels, laying the foundation both for progress in existing industries—as in the case of synthetic pesticides and fertilizers—and for the creation of entirely new ones, including plastics and synthetic fibers. Innovation in the information technology (IT) industry appears to have greatly increased the productivity of the U.S. economy: Jorgenson (2004) estimates that the quality-adjusted price of computers dropped at an annual average of 16 percent during 1959 to 2001 and that this rate of decline doubled after the mid-1990s, while Berndt and Rappaport (2001) report that personal computer prices declined an average of 35 percent *a year* between 1992 and 2002. The Internet has clearly changed all our lives, and the net effect of IT and telecommunications investment on national productivity appears to be quite high (Brynjolfsson and Hitt 2003; Jorgenson and Stiroh 2000). In the case of life sciences, major breakthroughs in areas such as heart disease, cancer, and Human Immuno-defiency Virus (HIV) have significantly reduced mortality rates (Cockburn 2007; Lichtenberg 2005; Duggan and Evans 2008). In each of these cases—with the possible exception of chemicals—well-funded, well-managed federal research laid the foundation for an industry that created great prosperity and that continues to be dominated by American firms.

The second reason why we believe that exploring the history of innovation in other industries may be useful is that it provides an intriguing perspective on some of the policy recommendations currently being advanced for the energy industry. Taken together, the histories point to three key factors as critical to accelerating innovation: (a) well-funded, carefully managed public research that is tightly linked to the private sector; (b) rapidly growing demand; and (c) antitrust, intellectual property, and standards policies that together promote vigorous competition and the entry of new firms. Expressed at this level of generality, these results echo the fundamental findings of an innovation policy literature that stretches back to Vannevar Bush (1945). Many scholars have noted the critical role played by well-managed federal research funds in shaping the United States' most innovative industries, and the idea that effective innovation feeds off both technology "supply" and market "demand" is a well-established one (Mowery and Rosenberg 1979). The innovation policy literature has focused less on the role of competition and new entrants in shaping innovation, but the intuition that new firms may have a critical role in spurring innovation goes back to Schumpeter (1934) and has recently been confirmed in a number of important studies (Aghion et al. 2005; Aghion et al. 2009). They are also broadly consistent with the bulk of the recommendations currently being made for accelerating innovation in energy.

But in innovation policy, the devil is in the details—particularly in the details of organizational implementation—and conclusions at this level of generality abstract from the critical institutional detail that is characteristic of each industry and that is particularly well captured in the chapters. The energy policy literature stresses the importance of adequate federal funding and the development of tight linkages between the public and private sectors (Anadon and Holdren 2009; Gallagher, Holdren, and Sagar 2006; Anadon, Gallagher, and Bunn 2009; Lester 2009). It explores the critically important role of designing policy instruments that will support demand for low-carbon technologies (Newell 2008; Nemet and Kammen 2007) and the potential role of intellectual property regimes in shaping innovation diffusion (Sagar and Anadon 2009). But, most important, it stresses the critical need for the development of a comprehensive innovation strategy that can coordinate policy across multiple entities and for the design of institutional structures that ensure public money does not become pork (Lester 2009; Ogden, Podesta, and Deutch 2008; National Commission on Energy Policy [NCEP] 2009; Weiss and Bonvillian 2009). In presenting the detailed histories of five industries where the delicate balance between public funding and private markets was maintained productively for many years, we hope to contribute in an important way to the conversation about exactly *how* these kinds of strategic and organizational choices can best be made.

Similarly, much of the current energy innovation policy literature stresses the potential importance of public funding of "deployment"—or of supporting the initial implementation of new energy technologies given their probable risk and scale. In four of the industries discussed here—agriculture, semiconductors, computers, and the Internet—the federal government played an important role in funding deployment of the new technology. Again, in presenting detailed case examples of the variety of ways in which the federal government filled this role, we hope to stimulate useful debate about what might be most appropriate in the case of energy.

A comparison between contemporary prescriptions for energy innovation and the history of accelerated innovation in a range of other industries is also intriguing because it highlights a number of cases in which the two are significantly different. The current energy policy literature, for example, focuses much more on innovation "supply" (public and private funding for research) and on innovation "demand" (the use of policy instruments to create demand for low-carbon energy) than it does on the instruments of competition policy. One of the striking findings from our industry histories, in contrast, is the (often seemingly unintended) role of antitrust and procurement policy in enabling widespread entry into the innovating industry. The histories also stress the important role that federally funded research has played in generating highly trained human capital for the private sector, an issue that has been stressed in some of the current energy innovation policy literature (Newell 2008, 2010), but not broadly so.

The remainder of this introduction frames the volume by attempting to draw together some of the key themes that occur across the chapters. The book then opens with Richard G. Newell's brief overview of the history of innovation in energy. The energy sector is vast and highly complex, but nonetheless Newell sketches out some "central tendencies" that provide important background for the chapters that follow. In chapter 2, Tiffany Shih and Brian Wright describe the history of innovation in agriculture, an industry that—in its time—had the scale and reach of the energy sector in our own day and in which both local funding for diffusion and federally funded innovation proved to be critical. Their chapter also raises a number of important issues around the role of intellectual property rights and public or private partnership in framing innovation. In chapter 3, Ashish Arora and Alfonso Gambardella explore the sources of innovation in the chemical industry, an industry that resembles some aspects of the energy sector in the chemical industry's focus on process development and whose early development was significantly more global than those of the others discussed here. Chapter 4 presents Iain M. Cockburn, Scott Stern, and Jack Zausner's review of the roots of innovation in life sciences, an industry that has been the second-largest recipient of federal research funding (after defense) and that is justly celebrated for the extraordinary productivity of the "innovation ecosystem" that links its public and private sectors.

Chapter 5 moves to David C. Mowery's review of the innovative history of the computer and semiconductor industries, and in chapter 6, Shane Greenstein discusses the institutional roots of the innovations that led to the development of the Internet. Given the centrality of information technology and telecommunications to the modern economy, it is particularly intriguing to be reminded of the critically important role that the federal government played in the development of both sectors—and to be exposed to careful discussions of the institutional structures and strategic decisions that made public support so very powerful in both industries. Finally, in chapter 7, Josh Lerner concludes the book with a detailed focus on venture capital, one often-cited potential solution to the clean energy innovation problem. He notes that despite the fact that venture capital investments are a relatively small share of total research and development (R&D) investment, they are exceedingly effective. However, he notes that while public policy can help to maintain a healthy level of venture capital investment, direct government investment in venture capital funding can exacerbate the boom and bust cycles characteristic of the industry.

Taken together, these chapters do not provide a comprehensive review of innovation policy because several excellent reviews of innovation policy exist (see, for example, Branscomb and Keller 1999; Cohen and Noll 1991; Mowery and Nelson 1999). Neither are they a comprehensive history of modern industrial innovation or of innovation policy. We were unable to persuade an expert on the history of innovation in the space or defense industries to participate in the project, for example, and with the notable exception of the chapters on agriculture and chemicals, the accounts by and large focus on the history of innovation in the United States. Mowery, Nelson, and Martin (2009) focus particularly on the history of innovation in the United Kingdom and on the relevance of the Manhattan and Apollo projects for our understanding of energy policy, and their paper is highly recommended. We believe, however, that given the centrality of the industries we explore to modern economic growth and the critical role of U.S. public policy in supporting each of them, our focus still yields important insights.

Last, we do not attempt to draw policy recommendations. This is in keeping with longstanding National Bureau of Economic Research (NBER) policy—but it also reflects our awareness that the energy sector is importantly different from the industries discussed here. We hope instead to provide the kind of rich data that will allow the interested reader to draw his or her own conclusions.

The Book's Key Themes

In broad outline, the chapters highlight three mechanisms that have historically served to support accelerated innovation: substantial, sustained, and effectively managed federal funding for fundamental research; the generation of growing customer demand, either through procurement or through the market; and the enabling of aggressive competition, particularly from newly entering firms.

Public Support for Fundamental Research

The hypothesis that one of the central roles of public policy with respect to innovation is to fund "basic" research is a well-established idea in the innovation policy literature (Bush 1945) and rests on the observation that the benefits of truly fundamental research are, in general, very difficult for private firms to appropriate. Much of the existing energy innovation policy literature stresses the central importance of this issue for U.S. innovation policy, stressing the relatively low levels of federal funding for fundamental energy research (Gallagher, Holdren, and Sager 2006; Nemet and Kammen 2007; Newell 2008; Ogden, Podesta, and Deutch 2008).

Publicly funded energy research constitutes about 3 percent of the total federal R&D budget (or less than 0.03 percent of gross domestic product [GDP]). Annual energy R&D funding decreased by more than half from a high of about $9 billion (real 2009 dollars) in the late 1970s to about $3 billion by the mid-1990s, where it remained for over a decade (see figure I.1; Newell 2008). This represented a drop from about 25 percent to 7 percent of nondefense federal R&D spending. Very recently, this trend has reversed: the allocation to energy research increased by over 20 percent, surpassing $4 billion in 2007, although this still represented less than 4 percent of total federal R&D spending. Energy R&D budgets increased modestly again in 2008, and in 2009 grew dramatically as a result of spending under the American Recovery and Reinvestment Act in response to the economic downturn. (Note that the Recovery Act spending includes significant amounts for technology demonstration projects.) While the Recovery Act spending represents a short-lived event, regular appropriations for energy R&D are slated to increase to about $6 billion by 2011, and the Obama administration has sought to further increase funding for basic research.

The most significant trend in recent years in federal R&D spending has been the large rise in defense R&D and in health-related R&D, which has increased from 25 percent of the federal nondefense R&D budget in 1980 to 54 percent in 2007 (see figure I.2). To place these figures in some perspective, in 2006, health expenditures accounted for 16 percent of GDP, energy

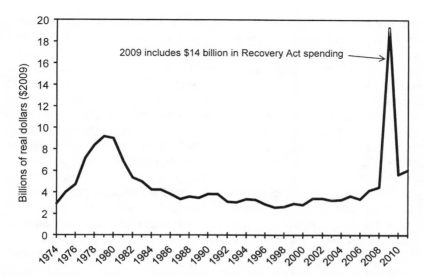

Fig. I.1   U.S. federal energy RD&D spending (1974–2009, with estimates for 2010–2011)

Sources: IEA (2010) U.S. Department of Energy (2010) for 2010 to 2011 estimates.

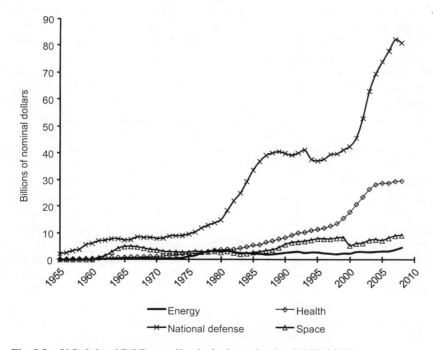

Fig. I.2   U.S. federal R&D spending by budget ufnction (1955–2008)

Source: NSF Science Indicators (2009, http://www.nsf.gov/statistics/fedbudget/).

expenditures accounted for 8 to 9 percent of GDP in recent years, and agriculture accounted for about 1 percent of GDP in 2004 (Newell 2008).

Faced with these numbers, many analysts have suggested very significantly increasing federal commitment to fundamental energy research (Gallagher, Holden, and Sager 2006; Nemet and Kammen 2007; Newell 2008; Ogden, Podesta, and Deutch 2008).

This idea is certainly consistent with the histories presented in our chapters. In every industry that we review, public support for fundamental research appears to have played a critical role in accelerating innovation in the industry, and there is some evidence that it has generated extraordinarily high returns. For example, Shih and Wright (chapter 2) suggest that marginal returns to public research in agriculture have been on the order of 45 to 60 percent. Similarly, Cockburn, Stern, and Zausner (chapter 4) highlight high, stable levels of public support as the primary foundation of industry's success in the life sciences, suggesting that this funding has led not only to many of the fundamental advances in scientific knowledge that have been indispensible to advances in modern molecular biology, but that it has also underwritten the development of a wide range of critical tools and techniques. For example, federal R&D underwrote the gene-splicing technique pioneered by Stanley Cohen and Herbert Boyer in 1973 that opened up the modern field of genetic engineering. Federal support for R&D was also critical to the early history of the semiconductor and computer industries.

For example, Mowery (chapter 5) notes that although the first solid state transistor was developed privately by AT&T, the firm's invention built on an extensive program of government-funded wartime research and that at the time it occurred, AT&T was in a competitive race with a publicly funded effort at Purdue University. Early work in the computer industry was entirely government funded, and from 1949 to 1959, federal funds were nearly 60 percent of computer-related R&D spending. Even as private spending grew more important on a percentage basis, federal funding continued to increase, with federally funded basic research in computer science growing from $65 million in 1976 to $265 million in nominal dollars in 1995. The early development of the Internet rested entirely on federal funding of the seemingly obscure technology of large-scale packet switching—a technology that was part of a portfolio of research whose commercial applications were entirely unknown at the time. Even in the case of chemicals—an industry in which many critical scientific discoveries were made in private firms—much of the foundational science of the field was performed inside universities.

Each of our authors suggests, however, that it is not only the magnitude of the public commitment to research funding that is critical, but also the ways in which this funding is governed. In this respect, the results we present here are very much consistent with those in the energy innovation policy community who have stressed that it is not enough for the federal government to dramatically increase funding for fundamental research: it is also critical that this research be managed appropriately (Lester 2009; Newell 2008; Ogden, Podesta, and Deutch 2008).

Four themes in particular emerge from the following chapters. First, in every industry, with the possible exception of chemicals, federal support for fundamental research was sustained over long periods of time. In both agriculture and life sciences, our authors explicitly point out that it took at least twenty years for fundamental research to have a notable effect on practice and that it is the "stock" of knowledge, not the flow, that has the greatest impact on accelerating private-sector innovation. For example, Shih and Wright note that it took several decades of development and learning before the U.S. land grant/State Agricultural Extension Service (SAES) system had acquired the scientific capacity and research base necessary to become an efficient system of innovation, and Cockburn, Scott, and Zausner suggest that the recent "surge and retreat" in National Institutes of Health (NIH) funding over the last decade has probably resulted in a less-productive innovation system and significantly distorted researcher incentives and career dynamics. In the case of IT, the groundbreaking investments were made in the 1940s and 1950s, but IT did not have a significant effect on the productivity of the economy until the 1990s. These examples suggest that a sustained investment in energy innovation may have groundbreaking implications— but that it may be years before these investments bear their full fruit.

Second, every author explores in some detail the ways in which effective governance steered a middle course between public funding of research that is so abstract and removed from potential application that it becomes isolated from practice and the danger that public funds become so applied that they will substitute for private funds. This is an old concern in the technology policy literature (Cohen and Noll 1991) and one that particularly concerns analysts of energy innovation policy (Newell, chapter 1 in this volume).

For example, Shih and Wright highlight the important role that locally supported, locally sited extension stations have played in ensuring that public funding for agriculture has been closely linked to actual practice. At the same time, they cite a 1986 analysis of a public/private research effort in Canada's malting barley industry in which the authors estimate that social gains would have been 40 percent higher if the public researchers had focused only on more fundamental research. Arora and Gambardella, in their study of the chemical industry, highlight ways in which tight links between U.S. universities and the private sector have been instrumental in codifying and diffusing scientific advances made by the industry.

In the case of the life sciences, Cockburn, Stern, and Zausner note that federally funded research is administered overwhelmingly through investigator initiated, peer-reviewed processes that are expected to ensure that the resources are allocated to genuinely promising scientific opportunities. Peer-review processes are not without their drawbacks, but they have the great advantage of putting every proposal through a rigorous "quality control" process controlled by experts in the field and, thus, of increasing the odds that the research will contribute meaningfully to the body of fundamental science. The authors stress, however, that the tight links that typically exist between public and private researchers in this sector serve to mitigate the risk that publicly funded research will become too removed from the problems and issues that are likely to have real-world application. Many publicly funded researchers work closely with private companies, and many privately funded researchers publish widely in the public literature and participate in academic conferences.

In the case of the Internet, in contrast, Greenstein stresses the unique role of Defense Advanced Research Project Agency (DARPA) in enabling fundamental, leading-edge research unencumbered by broader peer review. He focuses particular attention on DARPA's practice of funding "wild ducks"— creative researchers who were not constrained by institutional expectations. He also details the importance of involving a wide variety of researchers from many different backgrounds. Furthermore, program officers relied not only on their own deep knowledge of the technology in making critical funding decisions, but also on constantly bringing the network of researchers together to share ideas and to exchange criticism. They thus appear to have created a "market place for ideas" that was intensely robust and that generated very significant innovation. In both cases, the vast majority of publicly funded research was conducted through extramural performers, something that may also have contributed to the formation of effective links between research and practice. In general, the histories described in this volume highlight the fact that in each of these sectors, academic interaction with industry has been a two-way street—one in which important knowledge, expertise, people, and access to facilities have flowed in both directions.

Third, each of our authors stresses the important role that well-designed institutional mechanisms have played in effectively diffusing the results of federally funded work. In semiconductors and computers, for example, the federal government ensured the effective transfer of knowledge by requiring grant recipients to conduct seminars and distribute publications that actively disseminated information to others in the field. When AT&T developed the transistor, the military services that supported the company's transistor research also encouraged the dissemination of transistor technology. The proceedings of symposia funded by the military and held by Bell Labs (the research arm of AT&T) in the 1950s were widely distributed, and one produced two thick volumes on semiconductor technology that became known as "the Bible."

Greenstein's history of the Internet similarly provides a description of how DARPA's insistence that new technologies be embodied in prototypes and widely shared with other researchers led to the development of a powerful selection environment that mimicked the role of the market. Problems were quickly identified, and technologies that were found to be useful were immediately incorporated into other innovations, triggering a virtuous cycle of innovation that spread useful ideas rapidly around the research community.

Shih and Wright stress the critical role that the combination of SAESs and the federally funded agricultural extension service played in developing locally appropriate technologies and in ensuring their widespread diffusion. The authors explore the ways in which this structure balanced federal and state roles by combining federal financial support with state management of administration and direction of research. Shih and Wright suggest that the resulting structure provided an avenue to address local research needs while also exploiting interstate competition to motivate fruitful research.

The life sciences industry is also characterized by a particularly vibrant interchange between the public and private sectors. In their extensive discussion of the history and institutions that support this lively interface, Cockburn, Stern, and Zausner stress that the life sciences innovation network is highly decentralized and involves multiple linkages between and among different institutions, including universities, start-up firms, established biotechnology companies, pharmaceutical firms, government, and venture capitalists. The authors identify a number of factors as particularly important to maintaining this network. For example, they credit the combination of the Bayh-Dole Act, which allowed federally funded researchers to take title to their inventions and the *Diamond v Chakrabarty* decision, which established that genetically engineered organisms were eligible for patent protection, as instrumental in laying the foundations for the dynamic early-stage commercialization.

They also suggest that the creation of academic medical centers, which colocate publicly funded researchers working on fundamental problems with doctors who are actively treating patients, was a particularly important institutional invention. Although this structure was initially regarded with some suspicion, it has proved to be very fruitful—the authors suggest that it has led to numerous Nobel prizes and has led to numerous findings that were both fundamental scientific discoveries and also the basis for new commercially oriented technologies.

Several chapters in this volume explore the efficacy of public/private partnerships as mechanisms to ensure that publicly funded research is widely diffused. Here, no single lesson emerges. Arora and Gambardella, for example, explore the history of the synthetic rubber program in World War II. They suggest that the program met two of its initial objectives: it succeeded in greatly expanding the scale of synthetic rubber production and in improving synthetic rubber quality, and it attracted many new firms to the industry. It did not, however, make much progress toward its third goal, the generation and diffusion of significant new knowledge about polymers. Arora and Gambardella suggest that this may have been because the government insisted that participating firms adhere to a common recipe—thus slowing down the rate of innovation and leading the many research programs within the participating firms to be kept rigorously distinct from the publicly funded efforts.

In the case of agriculture, in contrast, Shih and Wright describe a number of cases in which public/private partnerships appear to have played an extremely positive role in both stimulating and diffusing critical knowledge. They note, for example, that research consortium models such as those adopted by the Latin American Maize Project (LAMP) and the Germplasm Enhancement of Maize Project (GEM) have been lauded for productively balancing public goods research with commercial viability. They also note, however, that while the U.S. Department of Agriculture has formed at least 700 cooperative research and development agreements with private firms, and while these have produced and commercialized some important innovations, researchers continue to be concerned that these arrangements will divert funds away from more basic research toward more applied work.

Last but not least, the research presented here shows that public funding of research can also play a critically important role in training the scientific and technical personnel who become the backbone of an innovative private sector.

Cockburn, Stern, and Zausner, for example, suggest that this dynamic has been fundamental to the development of the extraordinarily innovative private sector that characterizes the life sciences. They note, for example, that the practice of funding public research through peer-reviewed extramural NIH funding has created a high level of competition for funds and has supported the development of departments in universities focused on the biosciences. Within these departments, grant-supported training of PhD and postdoctoral students engaging in frontier research has helped to create a mobile, knowledge-based workforce that has moved fluidly between industry and academia. Between the early 1970s and today, the number of life science doctorate holders employed in academia more than doubled, and there has also been a significant expansion in the number of bachelor's-level students who receive a degree in the life sciences fields. Universities have played a similarly critical role in the chemical industry by institutionalizing the learning being created by firms and by training students. Arora and Gambardella credit universities for creating the disciplines of petroleum engineering, chemical engineering, polymer chemistry, and polymer engineering.

Mowery makes the point that federal support of fundamental R&D in semiconductors and computers played an instrumental role in building the "R&D infrastructure"—the institutions that identified and trained the highly skilled people whose work was fundamental to innovation in the commercial sector. For example, in the case of the software industry, the Semi-Automatic Ground Environment (SAGE) air defense project acted as a "university" of sorts for hundreds of software programmers—a development that laid the foundation for the future development of the industry within the United States. Similarly, the (much later) development of the so-called shrink-wrapped or mass market software industry was greatly aided by a (largely federally funded) university-based research complex. Greenstein makes an analogous point in the case of the Internet, suggesting that many of the key players who went on to private-sector careers developed their expertise in the context of early government-funded work.

Taken together, the work presented here is, thus, broadly consistent with much of the current energy innovation policy literature: in the highly innovative industries reviewed here, public funding for fundamental research was not only significant and sustained but also managed in such a way that it was tightly linked to practice, widely diffused, and led to the generation of a highly trained private-sector workforce. The ways in which this was accomplished, however, differ significantly across each industry, with each history highlighting governance mechanisms that may be worth considering in the energy context.

Demand and Induced Innovation

Rapidly growing demand plays two key roles in stimulating innovation. First and foremost, it signals a plausibly large and potentially rapidly growing market—something that greatly accelerates private-sector investment in innovation and the rapid diffusion of new technologies. Second, and perhaps more subtly, growing demand provides an important opportunity for immediate feedback from the market, whereby new product development underpins innovation that is more directly responsive to real market needs and less likely to fall prey to the isolation of the ivory tower.

In every one of the sectors explored here, rapidly growing demand triggered both extensive private-sector investment and extensive diffusion of new technology. For example, Cockburn, Stern, and Zausner's review of innovation in the life sciences suggests that one of the reasons R&D investment rates in the biopharmaceutical industry have remained so consistently high is because private firms have been consistently assured of robust demand for innovative products, as historically, so many health care needs have remained unmet. The authors note that the nature of demand for biopharmaceuticals has profoundly affected the life sciences innovation system. Intrinsically high willingness to pay for products that extend life or improve the quality of life—especially in the notably price-insensitive U.S. health care delivery system—has translated into relatively price-inelastic and stable demand. As a result, firms have been able to secure significant returns over a long period of time by focusing on the development and commercialization of innovative and novel biotherapeutic compounds. In agriculture, Shih and Wright find that similarly unmet needs—for higher yields and improved crop varieties able to withstand extreme weather, weed encroachment, and constantly evolving pests—have acted as a strong stimulus to innovation.

In both the semiconductor and computer industries, the large-scale entry of private firms coincided with a decline in prices, partly as a result of early government purchases, and an explosion in demand as both technologies allowed customers to do things that had never been done before. Similarly, the early commercialization of the Internet—and the excitement about its potential uses—was associated with a rush of private firms into the industry and a dramatic increase in the pace of innovation. Arora and Gambardella, in their chapter on the chemical industry, suggest that the explosion in demand for chemicals that accompanied the rapid industrialization of the early twentieth century was a critical factor in persuading private firms to invest heavily in chemical research. They also note that a lack of commercial demand was almost certainly the most important factor leading to the perceived failure of the government's Synfuels programs.

As many scholars have noted, one of the major barriers to the replication of these kinds of dynamics in the energy industry is the fact that "low-carbon" or "carbon-free" energy is typically indistinguishable from "dirty" energy at the point of consumption. In the industries whose history we include here—agriculture, chemicals, semiconductors, computers, the Internet, and biopharmaceuticals—once the technology reached some critical threshold of performance, demand exploded because the new technologies offered dramatic improvements over existing products, in many cases meeting consumer needs that had never been met before. This is unlikely to be the case with energy innovation.

Many authors working within the energy innovation policy literature have, therefore, argued that the single most important thing public policy can do to support the accelerated development and deployment of clean energy technologies is to create demand for low-carbon energy (Aghion et al. 2009; Anadon and Holdren 2009; Anadon, Gallagher, and Bunn 2009; Newell 2008, 2010; Stavins 2008; Weiss and Bonvillian 2009). Attention has focused on two kinds of mechanisms: on the creation of a "price" for carbon through some kind of tax or cap and trade regime and/or on the direct creation of markets through, for example, the imposition of renewable energy standards in energy purchasing or through the direct government support of first-in-class technology implementation through subsidy or purchase.

Proponents of the first approach stress its likely economic and technical efficiency (Stavins 2008; Newell 2010; Aghion et al. 2009). For example, Mowery, Nelson, and Martin (2009) suggest that a "Manhattan" or "Apollo" project approach is not an appropriate model for clean energy innovation.

In the defense and space industries, the government can entirely define the nature of customer demand because *it* is the final customer. In the case of clean energy, however, Mowery, Nelson, and Martin argue that it is deeply implausible to think that any discrete government program could foresee the precise technological solutions that will be appropriate. This is particularly true given that these technologies will need to be deployed throughout the world by many different actors, the deployment decisions will require huge outlays of private funds, and their largely embryonic state means that they will continue to evolve and improve over many years.

Proponents of the second approach, in contrast, stress the need for public subsidies to cross the "valley of death" between technological proof of concept and first commercialization, suggesting that this valley is likely to be particularly wide and expensive in the case of energy (Lester 2009; Ogden, Podesta, and Deutch 2008).

The results presented in the following chapters are consistent with both perspectives. On the one hand, there are several cases in which the federal government acted as the "first customer" for a new technology, arguably providing the critical support that was required to bridge the gap between prototype and private commercialization. Mowery's account of the early development of the semiconductor industry is an intriguing example of this dynamic in action. He documents how the prospect of large military procurement contracts in the early years of the semiconductor industry acted as a "prize," stimulating widespread entry and extensive innovation. He also details the important role that early military demand played in driving up industry production volumes and driving down production costs to the point where the new technology became commercially viable. He contrasts the success of these early efforts with the much more mixed track record of the Very High Speed Integrated Circuit (VHSIC) program—an early 1980s program that failed to meet its objectives and failed to successfully compete with the U.S. semiconductor market, which by then was dominated by commercial applications.

Mowery also documents the central role that federal procurement played in the early days of the computer industry. The first electronic U.S. digital computer was purchased by the military, and the first fully operational stored program computer built in the United States was purchased by the National Bureau of Standards. Even in the case of IBM's 650—the most commercially successful machine built in the 1950s—the projected sales of fifty machines to the federal government was critical to IBM's decision to move the computer to full-scale commercial development. In the 1970s and 1980s, the government's role as purchaser of high-performance computer equipment remained significant. Mowery further argues that federal procurement fostered the early development of the software industry. The rapid growth of the industry between 1969 and 1980 that gave the U.S. industry a worldwide advantage was spurred by federal willingness to invest in large, complex software development projects at a time when the commercial market for such projects did not exist.

Early government demand also looms large in Greenstein's account of how the DARPA-funded research that proved ultimately to be immensely useful despite the absence of immediate commercial demand. Greenstein makes the point that DARPA's investments in the development of large-scale packet switching and a "networks of networks" were considered highly risky and that no one foresaw any commercial application for them. This is not to say, however, that the early research went forward without any interaction with potential customers or without some sense of what demand might look like. Rather, the military had identified potential military uses for some of the new technology, and this potential military application shaped the early DARPA research.

Similarly, while it is the case that in none of the industries whose history is outlined in the following did the government explicitly set a price for the new technology, Cockburn, Stern, and Zausner suggest that in the case of the life sciences, the government has played a critical role in sustaining demand for innovative biopharmaceuticals. They hypothesize that—"whether by accident or design"—the interaction between the patent system, the Food and Drug Administration (FDA) regulatory process, and the way care is paid for within the U.S. health care system provide strong incentives for breakthrough innovation. The combination of a high willingness to pay for products, insurance that insulates purchasers from paying the marginal price, and the Hatch-Waxman regulatory framework provide strong incentives for the private sector to develop blockbuster therapies. These factors also provide incentives to develop a stream of innovations over time as the threat of generic entry upon patent expiration means that the returns to any single innovation are transitory.

What are the implications of these histories for our understanding of the role of government in creating demand for low-carbon energy? The industries we explore here differ crucially from energy in that one of the most important outcomes of innovation in each of them has been the creation of highly differentiated products that have met hitherto unmet needs, while "clean" energy cannot be easily differentiated at the point of delivery. It is thus difficult to imagine creating demand for low-carbon energy without government intervention—in the form of a price for GHG emissions, purchase mandates, or subsidies. Much ink has been spilled on the question of which of these interventions is likely to be the most economically efficient (see, for example, Popp, Newell, and Jaffe 2010; Jaffe, Newell, and Stavins 2005). The histories contained in this volume do not attempt to resolve this debate, but they do suggest that each may be effective, and they underscore the imperative of inducing clean energy demand if we are to see substantial, sustained, private-sector investment in energy innovation and the rapid deployment of new technologies.

Enabling Competition: Antitrust, Intellectual
Property, and Standards Policy

Accelerating innovation requires increasing both the supply of and the demand for new technologies. Beyond supply and demand, however, the theme that emerges most clearly from our histories is the important role that public policy has played in fostering vigorous competition and "markets for technology" in each industry and the centrally important role that this competition has played in accelerating innovation. Here again, our histories suggest that there is no single policy or set of policies that is always appropriate but that policy design must be actively tailored to the structure of the industry and the particular circumstances of the market. They focus attention on three policy instruments in particular: antitrust, intellectual property, and support for public open standards.

In the case of chemicals, Arora and Gambardella explore in some depth the role that appropriately narrowly defined patents and aggressive antitrust enforcement have played in encouraging entry in general and the rise of specialized engineering firms in particular. They note that new technologies in the industry have typically diffused extraordinarily fast, largely because of the presence of a robust market for technology fueled by the activities of small, independent specialized engineering firms. These firms not only build about 75 percent of all new plants, allowing easier entry into the industry, but are also responsible for about 35 percent of all new process inventions. Arora and Gambardella suggest that the existence of this market reflects the fact that patents in chemicals are both precise and narrowly specified. This makes the patents effective in protecting particular innovations and also allows competitors to enter the industry in closely related areas.

Arora and Gambardella also suggest that antitrust action has accelerated the widespread industry practice of licensing, focusing particularly on the cases of Standard Oil and DuPont. In 1909 to 1910, Standard Oil scientist William Burton developed the first commercially successful cracking process. It was a major innovation in refining technology, but Standard Oil was reluctant to invest in the process. However, as a result of an antitrust suit, the original Standard Oil was broken up into several firms in 1911, among which was Standard Oil of Indiana, where Burton worked. Standard Oil of Indiana not only commercialized Burton's process, but also licensed it to a number of other oil refiners. Similarly, DuPont was split into three separate firms following a successful antitrust suit in 1913 that also helped to convince DuPont's managers that the only path to future growth lay in entering new markets through innovation, rather than through the acquisition of existing producers.

In the life sciences, Cockburn, Stern, and Zausner note that there are multiple routes to the highly diversified, highly innovative private-sector "ecosystem" that characterizes the industry. The fact that this is one of the very few industries in which patent rights can be crisply defined appears to play an important role—notably by undergirding a vigorous biotechnology sector whose numerous small firms are largely venture capital funded. They note that the industry has seen the founding of more than 1,300 biotechnology companies in the United States and approximately 5,000 worldwide and that by the early 2000s, 25 to 40 percent of all pharmaceutical sales came from products having their origins in biotechnology. There has also been significant entry of specialized suppliers of biomedical materials and tools (e.g., gene sequencers and biomaterials); the development of contract research organizations that can provide expertise in areas such as early-stage clinical trials and the development of specialized managers, lawyers, and venture capitalists, who together can facilitate more effective transactions in what has become an increasingly complex web of relationships between academe, entrepreneurs, and downstream firms. Their analysis parallels Arora and Gambardella's suggestion that the emergence of a well-functioning "market for technology" can be hugely valuable in both stimulating private-sector innovation and in supporting its widespread diffusion. They stress the fact that both the structure of demand and the nature of regulation has meant that competition in the industry has been largely focused around innovation.

Greenstein's chapter describing the history of the Internet is one of the most eloquent on this set of issues. He suggests that effective competition was critical to the development of the Internet and rested on three key factors: "economic experiments, vigorous standards competition, and entrepreneurial invention." He stresses the crucial role that federal policy played in supporting all three factors. On the one hand, many of the established firms in the industry—including AT&T, the "Baby Bells," and IBM—actively rejected the possibility of investment in commercializing services related to Transmission Control Protocol/Internet Protocol (TCP/IP; the technological "core" of what later became the Internet) in the late 1980s. On the other hand, long-standing regulation of the telecommunications industry that favored new entry meant that the commercialization of the Internet was accompanied by a dramatic wave of new firm foundation.

Greenstein's account also explores the ways in which the process of standard setting in the industry was instrumental in supporting widespread, highly distributed innovation. Many of the key patents in the industry were publicly owned. This public ownership, coupled with the fact that early control of the technology rested largely in the hands of public-sector researchers, built a set of processes for standard setting that has been transparent and highly participatory. Greenstein makes the point that while this process has sometimes been frustrating, it has enabled the development of a highly complex value chain in which private, proprietary "platforms" coexist with public technologies in a way that makes it possible for small, innovative firms to innovate successfully without having to reinvent the entire system.

Mowery notes the importance to the semiconductor industry of both AT&T's ongoing antitrust litigation and of the government's procurement policies. Both encouraged the widespread diffusion of semiconductor technology and subsequent rapid entry into the industry. In the early 1950s, AT&T was reluctant, for antitrust reasons, to expand beyond its core base of telecommunications, thus leaving sales of the new technology into other applications as a tempting market for new entrants. At the same time, federal policies—driven partly by the desire to have "second sources" available for key military components—encouraged widespread diffusion of the new knowledge. For example, Mowery describes a symposium held at Bell Labs in 1951 attended by 130 industrial representatives, 121 military personnel, and 41 university scientists whose proceedings were widely distributed at government expense. The military was also willing to award large procurement contracts to newly founded firms, another mechanism that Mowery suggests was instrumental to the development of a highly competitive, highly innovative market structure that he contrasts with the very different semiconductor industries that emerged in Germany and Japan.

Mowery describes a similar dynamic in the case of computers—documenting how the military's belief that a strong technical infrastructure in support of innovation could only be built by the widest possible dissemination of technology—which led them to both use federal procurement policies to support new firms and to invest aggressively in information diffusion. He also describes how the IBM antitrust suit and subsequent consent decree was almost certainly instrumental in encouraging widespread entry into the hardware industry and the entry of many independent software vendors. He further observes that many of these entrants had been suppliers of computer services to federal government agencies. He speculates that U.S. import policies—which were notably more liberal than those adopted by Western European and Japanese governments—also played an important role in stimulating the competitiveness of the IT industry and the rapid declines in price-performance ratios that so accelerated the adoption of IT and the subsequent U.S. dominance of the industry.

In agriculture, in contrast, Shih and Wright voice deep concern over the role that patents play in the industry. They note the increasing evidence that multiple, mutually blocking intellectual property claims on inputs are hindering access to research tools that can be incorporated in the marketed products of agricultural research (Wright and Pardey 2006; Pardey et al. 2007). The authors suggest that the increasing concentration of the industry—for example, one estimate suggests that the top ten firms own more than half of all the agricultural biotech patents granted through 2000—is plausibly an attempt to retain "freedom to operate" by the major players and suggest that it may retard innovation in the industry. Even more critically, they suggest that the rising application of intellectual property rights to plant components and processes imposes high transaction costs for researchers who must acquire or license fragmented proprietary inputs to develop and commercialize a single downstream innovation. They further suggest that patents on locked-in but otherwise noncrucial genetic technologies have been retarding innovation and affecting the market structure of private research.

Although some energy policy analysts have explored the role of intellectual property regimes in shaping energy innovation (Reichman et al. 2008; Popp, Newell, and Jaffe 2010; Sagar and Anadon 2009), this has not been an area of central concern to scholars working in the field. Perhaps this approach reflects the belief that the scale of energy investment is such that only large established firms can be expected to introduce major innovations—but the importance of new entry to recent progress in both wind and solar technologies at least raises the question of whether some of the instruments that appear to have been important in the industries we study might also play an important role in stimulating vigorous, innovation-focused competition in the energy industry.

Conclusions

At the broadest level, the histories presented here are very much consistent with widely held views within the energy innovation policy literature. In general, this literature has suggested that greatly increasing rates of energy innovation requires creating significant demand for low-carbon technologies; substantially increased federal funding for well-managed research; and, in at least some cases, support for the initial deployment of new technologies. As the other markets explored in this volume do not face the same degree of unpriced environmental externality, there is no straightforward equivalent to a carbon price in the history of agriculture, chemicals, IT, or biopharmaceuticals. Nonetheless, our authors outline a number of ways in which public policy has often stimulated demand, particularly in the early stages of a technology's evolution, and confirm that the expectation of rapidly growing demand appears to have been a major stimulus to private-sector investment in innovation. Each history also confirms the centrality of publicly funded research to the generation of innovation, particularly in the early stages of an industry's history, and highlights a range of institutional mechanisms that have enabled it to be simultaneously pathbreaking and directly connected to industrial practice.

Our histories depart somewhat from the bulk of the energy innovation policy literature in focusing attention on the role of vigorous competition—particularly entry—in stimulating innovation, suggesting that in several industries, a mix of public policies—including procurement, antitrust, and intellectual property protection—played an important role in stimulating innovation by encouraging extensive competition and entry by newly founded firms. Many of the most innovative industries profiled here have been characterized by a lively "innovation ecosystem" that both rapidly incorporates the results of publicly funded research and supports widespread private-sector experimentation and rapid entry. There are, of course, important differences between the industries profiled here and the energy sector, but we believe that exploring the potential of these kinds of innovation ecosystems in clean energy might be a fruitful avenue for future research.

References

Aghion, Philippe, Nick Bloom, Richard Blundell, Rachel Griffith, and Peter Howitt. 2005. Competition and innovation: An inverted-U relationship. *Quarterly Journal of Economics* 120 (2): 701–28.

Aghion, Philippe, Richard Blundell, Rachell Griffith, Peter Howitt, and Susanne Prantl. 2009. The effects of entry on incumbent innovation and productivity. *Review of Economics and Statistics* 91 (1): 20–32.

Anadon, Laura Diaz, Kelly Sims Gallagher, and Mathew Bunn. 2009. DOE FY 2010 budget request and Recovery Act funding for energy research, development, demonstration, and deployment: Analysis and recommendations. Energy Technology Innovation Policy Group, Belfer Center for Science and International Affairs, John F. Kennedy School of Government, Harvard University, Working Paper.

Anadon, Laura Diaz, and John P. Holdren. 2009. Policy for energy technology innovation. In *Acting in time on energy policy,* ed. Kelly Gallagher, 89–127. Washington, DC: Brookings Institution.

Berndt, Ernst R., and Neal J. Rappaport. 2001. Price and quality of desktop and mobile personal computers: A quarter century historical overview. *American Economic Review* 91 (2): 268–73.

Branscomb, Lewis M., and James H. Keller. 1999. *Investing in innovation: Creating a research and innovation policy that works.* Cambridge, MA: MIT Press.

Brynjolfsson, Erik, and Lorin M. Hitt. 2003. Computing productivity: Firm-level evidence. *Review of Economics and Statistics* 85 (4): 793–808.

Bush, Vannevar. 1945. *Science, the endless frontier.* Report to the president. Washington, DC: GPO.

Clarke, Leon E., M. Wise, M. Placet, R. C. Izaurralde, J. P. Lurz, S. H. Kim, S. J. Smith, and A. M. Thomson. 2006. *Climate change mitigation: An analysis of advanced technology scenarios.* Richland, WA: Pacific Northwest National Laboratory.

Cockburn, Iain. 2007. Is the pharmaceutical industry in a productivity crisis? In *Innovation policy and the economy.* Vol. 7, ed. Adam B. Jaffe, Josh Lerner, and Scott Stern, 1–32. Cambridge, MA: MIT Press.

Cohen, Linda R., and Roger G. Noll. 1991. *The technology pork barrel.* Washington, DC: Brookings Institution.

Duggan, Mark G., and William N. Evans. 2008. Estimating the impact of medical innovation: A case study of HIV antiretroviral treatments. *Forum for Health Economics and Policy* 11 (2). http://www.bepress.com/fhep/11/2/1.

Edmonds, Jae, M. A. Wise, James J. Dooley, S. H. Kim, S. J. Smith, Paul J. Runci, L. E. Clarke, E. L. Malone, and G. M. Stokes. 2007. *Global energy technology strategy: Addressing climate change.* College Park, MD: Joint Global Change Research Institute, Battelle Pacific Northwest National Laboratory.

Energy Information Administration (EIA). 2010. *International energy outlook 2010.* No. DOE/EIA-0484(2010). Washington, DC: Energy Information Administration.

Gallagher, Kelly Sims, John P. Holdren, and Ambuj D. Sagar. 2006. Energy-technology innovation. *Annual Review of Environmental Resources* 31:193–237.

Grübler, Arnulf, Nebojša Nakićenović, and David Victor. 1999. Modeling technological change: Implications for the global environment. *Annual Review of Energy and the Environment* 24:545–69.

International Energy Agency (IEA). 2009. *World energy outlook 2009.* Paris: Organization for Economic Cooperation and Development/International Energy Agency.

———. 2010. *Energy technology RD&D budgets (2010 edition).* Paris: Organization for Economic Cooperation and Development/International Energy Agency.

Jaffe, Adam B., Richard G. Newell, and Robert N. Stavins. 2005. a tale of two market failures: Technology and environmental policy. *Ecological Economics* 54:164–74.

Jorgenson, Dale W. 2004. Productivity and growth: Alternative scenarios. In *Productivity and cyclicality in semiconductors,* ed. Dale W. Jorgenson and Charles W. Wessner, 55–59. Washington, DC: National Academies Press.

Jorgenson, Dale W., and Kevin J. Stiroh. 2000. Raising the speed limit: U.S. economic growth in the information age. *Brookings Papers on Economic Activity,* Issue no. 1:125–211. Washington, DC: Brookings Institution.

Lester, Richard. 2009. American's energy innovation problem. MIT-IPC-Energy Innovation Working Paper no. 09-007. http://web.mit.edu/ipc/research/energy/pdf/EIP_09-007.pdf.

Lichtenberg, Frank. 2005. The impact of new drug launches on longevity: Evidence from longitudinal disease-level data from 52 countries, 1982–2001. *International Journal of Health Care Finance and Economics* 5:47–73.

Manne, Alan, and Richard Richels. 1992. *Buying greenhouse insurance.* Cambridge, MA: MIT Press.

Mowery, David C., and Richard R. Nelson. 1999. Explaining industrial leadership. In *Sources of industrial leadership,* ed. D.C. Mowery and R. R. Nelson, 359–82. New York: Cambridge University Press.

Mowery, David C., Richard R. Nelson, and Ben R. Martin. 2009. Technology policy and global warming: Why new policy models are needed (or why putting new wine in old bottles won't work). Nesta, Working Paper. http://www.nesta.org.uk/library/documents/technology-policy-global-warming.pdf.

Mowery, David, and Nathan Rosenberg. 1979. The influence of market demand upon innovation: A critical review of some recent empirical studies. *Research Policy* 8:102–53.

Narayanamurti, Venakatesh, Laura D. Anadon, and Ambuj Sagar. 2009a. Institutions for energy innovation: A transformational challenge. Energy Technology Innovation Policy Research Group, Belfer Center for Science and International Affairs, John F. Kennedy School of Government, Harvard University. http://belfercenter.ksg.harvard.edu/publication/19572/institutions_for_energy_innovation.html.

———. 2009b. Transforming energy innovation. *Issues in Science and Technology* Fall:57–64.

National Commission on Energy Policy (NCEP). 2009. *Forging the climate consensus: The case for action.* NCEP Report. http://bipartisanpolicy.org/library/report/forging-climate-consensus-case-action.

Nemet, Gregory F., and Daniel M. Kammen. 2007. U.S. energy research and development: Declining investment, increasing need, and the feasibility of expansion. *Energy Policy* 35 (1): 746–55.

Newell, Richard G. 2008. A U.S. innovation strategy for climate change mitigation. Brookings Institution Discussion Paper no. 2008-15. Washington, DC: Brookings Institution. http://www.brookings.edu/papers/2008/12_climate_change_newell.aspx.

———. 2010. The role of markets and policies in delivering innovation for climate change mitigation. *Oxford Review of Economic Policy* 26 (2): 253–69.

Ogden, Peter, John Podesta, and John Deutch. 2008. A new strategy to spur energy innovation. *Issues in Science and Technology* Winter: http://www.issues.org/24.2/ogden.html.

Pardey, Philip G., Jennifer James, Julian Alston, Stanley Wood, Bonwoo Koo, Eran Binenbaum, Terrence Hurley, and Paul Glewwe. 2007. *Science, technology, and skills.* St. Paul, MN: International Science and Technology Practice and Policy (INSTePP).

Popp, David, Richard G. Newell, and Adam B. Jaffe. 2010. Energy, the environment, and technological change. In *Handbook of the economics of innovation.* Vol. 2, ed. Bronwyn Hall and Nathan Rosenberg, 873–937. Amsterdam: Elsevier B. V.

Reichman, Jerome, Arti K. Rai, Richard G. Newell, and Jonathan B. Weiner. 2008. Intellectual property and alternatives: Strategies for a green revolution. Energy, Environment, and Development Programme Paper no. 08/03. London: Chatham House.

Sagar, A. D., and L. D. Anadon. 2009. Climate change: IPR and technology transfer. Working Paper. Bonn, Germany: United National Framework Convention on Climate Change (UNFCCC) Secretariat, September.

Schumpeter, Joseph A. 1934. *The theory of economic development.* Cambridge, MA: Harvard University Press.

Stavins, Robert. 2008. Addressing climate change with a comprehensive U.S. cap and trade system. *Oxford Review of Economic Policy* 24 (2): 298–321.

U.S. Department of Energy. 2010. FY 2011 congressional budget request: Budget highlights. Washington, DC: U.S. Department of Energy.

Weiss, Charles, and William Bonvillian. 2009. *Structuring an energy technology revolution.* Cambridge, MA: MIT Press.

Wright, Brian D., and Philip G. Pardey. 2006. Changing intellectual property regimes: Implications for developing country agriculture. *International Journal of Technology and Globalization* 2 (1-2): 93–114.

1

The Energy Innovation System
A Historical Perspective

Richard G. Newell

While the importance of innovation in the energy technology arena is widely understood—particularly in the context of difficult problems like climate change—there is considerable debate about the specific role of public policies and public funding vis-à-vis the private sector. To what extent can the market drive innovation in new, lower-carbon energy technologies once regulatory constraints have been adopted and prices begin to capture the environmental externality associated with greenhouse gas (GHG) emissions? Accepting that a rationale exists for direct public research and development (R&D) investment even in the context of a pricing policy, how much investment is justified, and what mechanisms and institutions would most effectively deliver desired results? What lessons can be drawn from the past thirty years of federal involvement in energy technology R&D, and what do they imply about government's ability to pursue particular energy-related policy objectives?

These questions are important precisely because the potential economic payoff from well-designed policies is high, with annualized cost savings from advanced low- and no-GHG technologies being estimated in the tens to hundreds of billions of dollars per year (Newell 2008). At the same time, public resources are likely to be substantially constrained going forward given the current long-term fiscal outlook in the United States and elsewhere. This

Richard G. Newell is the Gendell Associate Professor of Energy and Environmental Economics at Duke University's Nicholas School of the Environment, and university fellow of Resources for the Future. He currently serves as administrator of the U.S. Energy Information Administration (EIA), and was a research associate of the NBER prior to the start of his government service.

Special thanks to Marika Tatsutani for exceptional assistance and Rebecca Henderson for comments on the chapter.

reality prompts additional questions: first, what options realistically exist for funding expanded investments in energy technology innovation? Second, what institutions are best positioned to direct and oversee publicly funded technology programs?

1.1   Highlights from the History of Energy Innovation

Technological innovation in the production and use of energy is inextricably interwoven with the larger history of human development—indeed, the ability to harness ever larger quantities of energy with ever increasing efficiency has been central to, and inseparable from, the improvements in living standards and economic prosperity achieved in most parts of the world since pre-Industrial times. Sketched in broad terms, progress has been dramatic. According to a recent report by the United Nations Development Program (UNDP), for example, the simple progression from sole reliance on human power to the use of draft animals, the water wheel, and, finally, the steam engine increased the power available to human societies by roughly 600-fold (UNDP 2000). The advent of the steam engine, in particular, had a transformative effect, making the production of energy geographically independent of proximity to a particular energy source (because the coal used to power steam engines could be transported more or less anywhere) and ushering in the Industrial Age.

In the decades that followed, advances in energy technology continued and even accelerated, often with far-reaching implications for day-to-day aspects of human life, especially in the world's industrialized economies. The electrical grid and other major system innovations were introduced, and individual technologies continued to improve. Ausubel and Marchetti (1996), for example, estimate that the efficiency of steam engines improved by a factor of roughly 50 since the 1700s; modern lighting devices, meanwhile, are as much as 500 times more efficient than their primitive forebears. As available means of producing and using energy became more convenient, portable, versatile, and efficient, overall demand also increased: citizens of developed countries now routinely consume as much as 100 times the energy their pre-Industrial ancestors did (UNDP 2000).

Additional compelling evidence for continued innovation in the energy realm can be found in broad macroeconomic indicators—most notably in the fact that the amount of energy required to produce a unit of goods and services in the world's industrialized economies has declined steadily since the mid-1970s. According to various estimates, the energy intensity of the United States and other Organization for Economic Cooperation and Development (OECD) countries has been falling by approximately 1.1 percent per year over the last three decades. Importantly, similar trends also began emerging in a number of major non-OECD economies (such as China) in the 1990s as these countries began to modernize from a relatively inefficient industrial base (UNDP 2000). As a result, the world as a whole now produces more wealth per unit of energy than ever before.

While these broad trends can be documented with relative ease, the specific role of innovation per se—as distinct from investment, learning during use, structural change in the economy, and other factors—is much harder to quantify. In part, this is because the energy sector itself is unusually large, diverse, and complex. There are numerous distinct technologies and industries for producing and converting primary sources of energy, such as petroleum, coal, and natural gas extraction and combustion; nuclear, hydroelectric, solar, and wind power; as well as biofuels. At the same time, there has also been significant investment in the technologies of energy distribution—such as the electrical grid and pipelines—and, perhaps even more critically, in the technologies of energy use, which include everything from home appliances to automobiles and office equipment. Entire books or reports have been written on innovation in each of these areas alone; undertaking an authoritative treatment of the subject for energy broadly defined would be extremely challenging, to say the least.

Given the inherent difficulty of generalizing over such a broad and diverse set of technologies and industries, we focus in the next section on the record of innovation over the last half century or so in a few key areas: conventional energy resources, primarily oil, coal, gas, and nuclear; renewable energy technologies, primarily wind and solar; end-use energy efficiency; and pollution control. In all cases, we provide at most a brief review; a more extensive literature can be accessed through the sources cited here. Despite the limitations of this necessarily cursory overview, however, a few important themes or insights emerge:

1. Viewed from the standpoint of historic improvements in the efficiency of energy resource extraction and use, there are grounds for substantial optimism about the innovative potential of energy technology industries.

2. From the standpoint of efforts within the last half century to develop wholly new energy supply options and, in particular, to reduce humanity's reliance on conventional fossil fuels, however, the record is far more mixed. With the possible exception of civilian nuclear power, which developed as a by-product of R&D investments undertaken for military purposes, substantial public investments in alternative energy have by and large not yielded game-changing technological advances that would allow for a fundamental shift in the distribution of primary energy sources.

3. Where there is no market demand (or "pull") for a particular energy technology improvement, the investment of public resources to "push" innovation has typically yielded poor returns. In energy, markets for new technologies have usually emerged when one (or more) of the following occurs: (a) prices for conventional resources rise as a result of rising demand and stagnant or falling supply or production capacity; (b) technological possi-

bilities arise that more effectively meet energy demands; and (c) government imposes new policies or regulations that affect market conditions for energy technologies. Classic examples of the latter would include pollution control requirements, efficiency standards, technology mandates (such as renewable portfolio standards), or technology incentives (like the renewable energy production tax credit).

4. To the extent that markets for new energy technologies greatly depend on public policies or public funding, they are inherently vulnerable to fluctuations in political support. Uncertainty about the future continuity of policies or funding can discourage private-sector investment and create boom-bust cycles for new energy technologies (examples of this dynamic can be found in the history of several renewable energy industries and in the U.S. synfuels program of the late 1970s and 1980s).

1.2   The Record of Innovation in Energy Technology: A Brief Review

1.2.1   Fossil Fuels

Fossil fuels—coal, oil, and natural gas—today supply over 80 percent of the world's energy needs. Decades of incremental technology improvements have led to major productivity gains in the extraction and processing of these resources. For example, U.S. miners in 1949 produced 0.7 short tons of coal per miner hour; fifty years later, the rate was over 6 short tons per miner hour (EIA 2009a). Similarly, dramatic advances have occurred in the oil industry, which continues to improve the technology for locating and extracting new reserves. As a result, estimates of the remaining recoverable petroleum resource base are continually being revised upward, despite high rates of global consumption and periodic concerns about dwindling global supply.

For example, in 2000, the U.S. Geological Survey (USGS) estimated ultimately recoverable reserves of conventional oil at 3.3 trillion barrels worldwide (including natural gas liquids), of which roughly one-fifth had already been produced at that time (USGS 2000). Taking into account improvements in seismic tools, imaging software and modeling tools, and new extraction techniques (such as the use of horizontal wells), the consulting group Cambridge Energy Research Associates (CERA; 2006) estimated global recoverable reserves at as much as 4.8 trillion barrels. Advanced secondary and tertiary recovery technologies have also made it possible to extract more oil from existing fields. According to the *New York Times,* Chevron estimates that it can recover up to 80 percent of the oil at an existing field near Bakersfield, California, using advanced recovery techniques; originally, the company had estimated it could recover only 10 percent of the oil at this site (the industry average is approximately 35 percent; (Mouawad 2007). Similar trends exist in natural gas extraction, with recent advances in gas shale significantly expanding U.S. gas resources.

The record of improvement in major fossil-fuel-based conversion technologies, by contrast, is more mixed. On the one hand, the typical thermal efficiency of conventional, steam-electric, coal-fired power plants has remained relatively unchanged for decades at 30 to 40 percent (InterAcademy Council 2007). More-recent innovations, such as fluidized bed or supercritical coal systems can boost generation efficiency and reduce emissions of key air pollutants, but these technologies—while commercially available and already in use at a number of facilities around the world—have been slow to achieve significant levels of market penetration. This is in large part because the rate of turnover of old coal plants and the construction of new plants in developed countries has been quite slow in recent years, while the cost of more-advanced systems remains a major impediment in the developing or emerging economies that have been adding coal capacity more rapidly. Gasified coal systems, which hold out the promise of facilitating further efficiency gains as well as cost-effective carbon capture, remain relatively untested at a commercial scale—in part because they face formidable deployment hurdles.[1]

Thus, the most important efficiency gains in electricity generation in modern times have been achieved through the introduction of advanced, combined-cycle turbines that operate on natural gas. These types of systems have dominated new capacity additions in the United States and elsewhere for more than a decade, in large part because they have low pollutant emissions and can be built quickly, on a smaller scale, and at lower capital cost than other power options.

A similarly mixed picture applies to the major existing conversion technology for petroleum used in transportation applications: the internal combustion engine. On the one hand, engineering improvements have substantially boosted the output of power from such engines per unit of fuel input. On the other hand, the extent to which engine efficiency improvements have translated into improved fuel economy (as opposed to increased power or vehicle size and weight) has depended highly on fuel prices and government policies. In the United States, a boost in vehicle efficiency standards after the oil crisis of the 1970s was followed by a long period of stagnation in overall fuel economy after the mid-1980s. In Europe, by contrast, high fuel taxes and other factors have led to a higher-mileage auto fleet. In the last several years, U.S. policy and market trends have again shifted toward higher fuel economy.

Efforts to develop alternative transportation fuels, meanwhile, have produced some of the most problematic examples of U.S. energy policy to date. In particular, the launching of the Synfuels Corporation in 1980 repre-

---

1. Although the component technologies involved in gasification systems have been widely used in the chemical and refinery industries for decades, they have not been widely demonstrated at a commercial scale for electric power production. Thus, the technology is perceived as more costly and more risky by the electric power industry, and first-mover projects have had difficulty attracting sufficient private-sector or utility investment.

sented the culmination of a multiyear, multibillion-dollar U.S. Department of Energy (DOE) effort to develop methods for producing petroleum from unconventional domestic sources such as coal or oil shale. The effort collapsed without achieving its major objectives in 1986 following a substantial decline in oil prices. A more recent focus on the development of biomass-based alternative transportation fuels has produced a rapid and dramatic expansion of ethanol production in some parts of the world, notably the United States and Brazil. However, significant technology advances involving the utilization of new feedstocks or conversion technologies that could dramatically reduce the cost, energy, and environmental requirements of biofuels production remain for the most part in the precommercial, research, development, and demonstration (RD&D) phases of development.

1.2.2   Nuclear

Against this backdrop, nuclear power offers perhaps the most dramatic example of a major energy supply innovation that was deployed on a large scale within the last half century. Developed as an outgrowth of military R&D investments, civilian nuclear power experienced a relatively brief period of substantial commercial investment from the 1970s to the mid-1980s, based on the hope—especially compelling in the immediate aftermath of the 1973 oil crisis—that it might eventually provide a near-limitless, domestic supply of energy at a price that was "too cheap to meter." In a time span of fewer than two decades, nuclear power grew to contribute roughly 16 percent of global and 20 percent of U.S. electricity supply (in a few countries, such as France, it accounts for a significantly larger share; World Nuclear Association 2005; EIA 2009c).

Since the 1980s, however, further nuclear capacity additions have slowed dramatically due to a combination of high capital costs relative to other conventional generation options and concerns about a range of related issues, from waste management to weapons proliferation and public safety—concerns that were heightened in the wake of widely publicized accidents at Three Mile Island in 1979 and Chernobyl in 1986. Nevertheless, the nuclear industry worldwide has been able to maintain a roughly stable share of overall electricity supply, in large part because of ongoing improvements in the operating efficiency of existing plants. In fact, the average utilization or "capacity factor" of U.S. nuclear plants increased from 56 percent in 1980 to 66 percent in 1990 and over 90 percent currently (EIA 2009d).

Despite at best uncertain prospects for a second wave of nuclear power plant construction, governments around the world never stopped investing in the technology, which has continued to evolve through several generations of new designs. Most reactors operating today are considered Generation II; more recent reactors built in France and Japan utilize Generation III designs, which emerged in the 1990s with the idea of reducing costs through increased standardization and other innovations. Generation III+ designs incorporate further improvements, including passive emergency cooling systems in place of conventional power-driven systems. In 2002, ten nations and the European Union launched a coordinated R&D effort, known as the Generation IV International Forum (GIF), to develop a new set of reactor designs that take advantage of high-temperature, high-efficiency concepts to substantially reduce waste output and fuel use. Participants in the GIF are pursuing focused research on six different types of reactor designs, including the very high temperature gas reactor, the supercritical water reactor, the lead-cooled fast reactor, the sodium-cooled fast reactor, the gas-cooled fast reactor, and the molten salt reactor.

Continued rapid growth in global electricity demand together with mounting concerns about climate change led to a widespread perception earlier this decade that the nuclear industry could be poised for a second major wave of expansion. Bolstering that perception, a number of new units utilizing recent technology or design innovations have been proposed in the United States and elsewhere in the last several years, even as a number of governments introduced or strengthened existing policies and subsidies—including loan guarantees or other incentives—to support new plant construction. More recently, however, construction cost increases across many large-scale engineered projects, a worldwide economic slowdown, and actual experience with the construction of a new reactors in Finland and France may have dampened prospects for a renaissance of the civilian nuclear power industry (Deutch et al. 2009).

1.2.3   Renewables

Renewable energy has been another area of major public- and private-sector investment in new energy supply options—one that like nuclear power and synthetic fuels had its roots in the post-oil embargo era of the late 1970s and early 1980s. In the 1970s, a number of countries began a major push to develop wind and solar technology; early R&D efforts in the United States were funded by the federal government, along with the National Aeronautics and Space Administration (NASA) and Boeing. Efforts were soon bolstered by the introduction of generous tax incentives. These efforts led to a "wind rush" in the early 1980s that saw the construction of the first large-scale wind farms, mostly in California. Denmark also made an early and substantial investment in wind, emerging as a leader in the production and design of wind turbines by the 1980s. In the United States, the locus of innovative activity increasingly shifted to a number of smaller entrepreneurs who continued tinkering with different rotor and gearbox designs even as the commercial wind industry ground to an abrupt halt in the mid-1980s, when state and federal tax credits began to expire (see *Economist* (2008) for an overview of the history of wind technology development and Neij (1999, 2005) for a discussion of the cost dynamics of wind power).

With the benefit of the design improvements that emerged from these efforts and those of the Danish manufacturers, wind investment in the United States took off again in the early 2000s, propelled by the reintroduction of tax credits and a growing number of prorenewable state policies. Recent years have seen dramatic worldwide growth in installed wind capacity, which rose from 18 gigawatts in 2000 to a global total of 159 gigawatts by the end of 2009—a trend that is projected to continue into the future (EIA 2010). Before the current economic downturn, in fact, some analysts were predicting that wind would grow to as much as 2.7 percent of global electricity generation by 2012 and nearly 6 percent by 2017 (*Economist* 2008). Although under current policies, EIA (2010) projects more modest growth to a 2.3 percent share by 2015 and 3.6 percent share by 2020, the rate of growth is still almost 14 percent per year.

Meanwhile, wind technology itself has also undergone substantial changes: early wind turbines tended to be relative small, with generating capacities on the order of tens of kilowatts and rotor diameters on the order of 15 meters. More recent turbines benefit from the ability to operate at variable speeds and use lighter-weight materials; this has allowed the introduction of much larger units, which in turn has produced substantial cost reductions. Wind turbines built in recent years typically generate 1.5 to 2.5 megawatts and have rotor diameters as large as 100 meters; recent proposals have featured even larger turbines. The per-kilowatt-hour cost of generating electricity from wind, meanwhile, has fallen from an industry average of thirty cents in the early 1980s to approximately ten cents in 2007 (*Economist* 2008).

As this brief review suggests, the development of wind and other new energy technologies has been strongly influenced by financial incentives and other policy support from the public sector.[2] Federal tax incentives—for electricity production in the case of wind and for investment in the case of solar—were particularly critical drivers of deployment and innovation for these technologies. The current federal renewable energy production tax credit dates back to the Energy Policy Act of 1992, which provided a 1.5 cent-per-kilowatt-hour tax credit for the first ten years of power output from qualifying wind and biomass facilities. The tax credit was indexed to inflation and now totals 2.1 cents per kilowatt-hour. Since its inception, the production tax credit has been extended or renewed multiple times, but always for periods of at most two to three years at a time. Moreover, on five occasions since 1999, the program has actually expired before being renewed, often with some changes in eligibility requirements and other rules.

---

2. Tax credits and other incentives have also been used to promote energy technologies other than wind and solar. In the United States, for example, production tax credits have also been available for advanced coal and nuclear power. Other prominent examples of energy technology subsidies in the U.S. context include the excise tax credit for ethanol, liability protection for the nuclear industry in the form of the Price-Anderson Act, and federal loan guarantees for the construction of new nuclear power plants.

This pattern has created substantial investment uncertainty for the industry: in years when tax credits lapsed, capacity additions fell precipitously compared to the prior year.

Solar energy, meanwhile, has historically benefited from a 10 percent investment tax credit, although it was also eligible for the production tax credit for a brief period from 2004 through 2005. Under the Energy Policy Act of 2005 and subsequent reauthorizations, the investment tax credit for solar energy increased to 30 percent of eligible system costs. Overall, solar technology has yet to achieve the level of cost-competitiveness and market penetration of wind—especially in centralized, grid-connected applications—but the solar industry has likewise experienced dramatic global growth in recent years and achieved significant cost reductions (Watanabe, Wakabayashi, and Miyazawa 2000).[3] Earlier this decade, the solar energy industry as a whole—which includes solar thermal and photovoltaic (PV) technologies in both grid-connected and stand-alone applications—experienced average annual growth rates in excess of 40 percent (DOE 2009). Installed PV capacity, most of it grid-connected, grew especially quickly to a cumulative global total of more than 16 gigawatts (peak capacity) by the end of 2008 (REN21 2009). Meanwhile, the best commercially available PV cells now achieve conversion efficiencies above 23 percent, well above the current industry average of 12 to 18 percent (EIA 2010). Even higher efficiencies—in excess of 40 percent (NREL 2008)—have been achieved in the laboratory. By comparison, the conversion efficiency of the first solar cell developed by Bell Laboratories in 1954 was 6 percent (EIA 2010).

Despite this progress, however, remaining cost and deployment hurdles for solar are such that the industry's commercial prospects going forward will continue to depend strongly on government support, including both direct support in the form of financial incentives and public R&D investments and indirect support in the form of GHG regulation and other public policies designed to advance renewable or alternative energy sources.[4] With average levelized electricity production costs on the order of twenty-five cents per kilowatt-hour (EIA 2010), solar PV remains substantially more expensive at present than competing conventional power options and, like wind, it faces challenges related to siting, intermittency, and grid integration.

---

3. Much of the recent demand for solar technology has come from decentralized, stand-alone applications—including rooftop installations and as a power source in remote locations or developing-country settings.

4. An important deployment hurdle for both wind and solar technology is the availability of adequate transmission infrastructure, particularly to relatively remote sites where the underlying resource potential tends to be more concentrated. Continued advances in grid technology and capacity are also critical to support renewable energy technologies whose output—in contrast to conventional power sources—varies according to weather conditions and time of day.

1.2.4   Energy Efficiency

A rich and far-ranging record of technology innovation can also be found on the demand side of the energy equation, in the evolution of the wide variety of devices and appliances that use energy to do work and provide light, heat, refrigeration, mobility, air conditioning, and a host of other services and amenities. Although the topic of innovation in energy efficiency is more extensive than can be summarized adequately here, it is worth noting that public R&D investments in this area, according to at least one relatively recent study of the past record of DOE programs in the United States, have yielded far larger economic cost savings and other societal benefits than past public investments in fossil supply technologies (National Research Council [NRC] 2001). Energy efficiency advances also provide numerous examples of the interaction between innovation and regulatory policy in accelerating innovative progress.

The case of refrigerator technology, for example, has been frequently cited because it dramatically illustrates the potency of these interactions. In the United States in the early 1990s, publicly supported R&D efforts combined with innovative utility programs led to significant improvements in refrigerator and freezer technology. These improvements led to the enactment of state and eventually federal minimum efficiency standards for refrigerators, motivating further innovation and continued technology advances as the standards became more stringent in subsequent years. The resulting marketwide improvement in refrigerator and freezer efficiency has been credited with producing very substantial and highly cost-effective cumulative reductions in energy consumption over a period of multiple years. The average refrigerator today consumes 75 percent less energy than its 1975 counterpart, even though it typically has larger storage capacity, more features, and costs less in inflation-adjusted terms.

Similar examples of innovative progress can be found in other energy end-use technologies and in energy-intensive industries, such as steel and cement manufacturing, which face strong private incentives to improve energy efficiency as a means of enhancing overall cost-competitiveness. For example, according to figures compiled by the U.S. EIA, the average energy intensity of the U.S. iron and steel industry—as measured by the first use of energy for all purposes in thousand Btu divided by the value of production in constant 1992 dollars—declined by more than 25 percent in a single decade from the mid-1980s to the mid-1990s (from 46.47 thousand Btu per dollar in 1985 to 33.98 thousand Btu per dollar in 1994; EIA 2006). Moreover, data collected by EIA in subsequent years show that the energy intensity of the U.S. iron and steel industry continued to decline between 1998 and 2002. Research by Popp (2001) using patent data from thirteen energy-intensive industries suggests that investments in efficiency technologies by these industries have generally been highly cost-effective. Specifically, Popp finds that the median patent leads to $14.5 million dollars in long-run energy savings, while the industries that use these technologies spent an average of $2.25 million of R&D per patent.

1.2.5   Pollution Control

A final area of energy technology that has been studied for evidence of its effects on innovation concerns pollution control. Here, too, numerous examples can be found where dramatic advances were achieved in technology performance and cost across multiple industries and types of pollution. In most cases, these improvements were prompted by the introduction of mandatory regulation given the public good nature of pollution reductions. When limits on sulfur dioxide (SO_2) emissions from power plants were being debated in the United States in the late 1980s, for example, government and industry estimates indicated that the costs of pollution abatement would likely be on the order of $1,000 per ton or more. Under the market-based Acid Rain Program that was eventually introduced, however, abatement costs proved dramatically lower than expected. Indeed, SO_2 allowance prices throughout the first decade of program implementation remained fairly stable at or below $200 per ton (EPA 2009).[5]

In fact, a number of studies have looked at the effects of innovation on the costs of pollution abatement as one measure—albeit an incomplete one—of returns to R&D investment. For example, Carlson et al. (2000) examine changes in the marginal abatement costs for air pollutant emissions at power plants and find that about 20 percent of the change in marginal abatement costs that have occurred from 1985 to 1995 can be attributed to technological change. Popp (2003) uses patent data to link innovative activity to lower operating costs of scrubbers for coal-fired electric power plants. He finds that a single patent provides a present value of $6 million in cost savings across the industry. Assuming approximately $1.5 million of R&D spent per patent granted, this yields a rate of return similar to those found in the more general technological change literature.

1.3   Drivers of Energy Technology Innovation: The Role of Markets and Government Policy

Historically, a number of market and regulatory conditions have influenced private- and public-sector spending on energy-related R&D. Trends over the last half century suggest that investment tends to decline when energy prices are low and when available production capacity and technol-

---

5. The flexible, market-based structure of the cap-and-trade regulatory approach used in this instance is widely credited with producing these cost reductions (see, for example, Stavins 1998). Note that SO_2 allowance prices began to move upward in 2005 in anticipation of further federal regulations; they remained high relative to historic levels in 2006 and 2007. By mid-2008, however, allowance prices had again fallen to below $200 per ton.

ogies are perceived to be ample, or at least adequate to meet market demand. When prices rise because of a perception of resource scarcity or because government policies—in the form of changed regulation or incentives—create a shift in market conditions, investment tends to increase. Following the Organization of the Petroleum Exporting Countries (OPEC) oil embargo of 1973, for example, energy prices rose sharply, and governments around the world instituted policies aimed at reducing dependence on imported oil. As a result, investments in energy-related R&D—by both the public and private sectors—grew rapidly, reaching a historic peak roughly around 1980. Subsequent spending, however, declined substantially in real terms, reflecting the fact that fossil-fuel prices were low for most of the 1980s and 1990s, along with market structure changes in the power industry (Sanyal and Cohen 2009). The trend of falling expenditures on energy R&D during this period was compounded in the United States by the deregulation or restructuring of the natural gas and electric utilities industries and efforts to balance the federal budget.

A more recent shift in market and regulatory conditions for energy technology occurred earlier this decade when oil and natural gas prices began to climb in response to rapidly growing global demand, and governments began introducing policies motivated by a new set of environmental and energy security concerns. The result was a resurgence of public and private investment in energy-related R&D and rapid growth in some alternative energy industries, such as wind and biofuels. These trends have recently been complicated by the global economic slowdown and stresses within financial markets that began in 2008. The full impacts of the current crisis are not yet clear. On the one hand, an abrupt slackening of global demand led to a marked drop in energy prices, while tight credit markets have created new barriers to investment. On the other hand, economic stimulus efforts in the United States and elsewhere are contributing—at least in the short run—to increased investment in alternative energy sources and efficiency improvements. Energy prices have also advanced from their recent lows.

Historic shifts in public funding for energy R&D, both in terms of the overall level of spending and in terms of the emphasis on different types of resources, are illustrated by figure 1.1, which shows spending by the U.S. DOE on energy R&D. The figure indicates that current expenditures now total more than $5 billion annually. This represents a marked increase over funding levels at the start of this decade, but it remains about half, in inflation-adjusted terms, of the peak level of spending reached in 1979.

Data on energy-related R&D spending by private firms are more difficult to obtain. Broad estimates suggest that direct federal spending—which cumulatively totaled more than $100 billion in real terms over the last three decades (most of it spent through DOE programs)—represented about one-third of total national expenditures on energy R&D, with the balance being

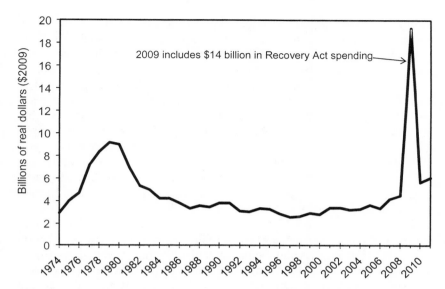

Fig. 1.1   U.S. federal energy RD&D spending (1974–2009, with estimates for 2010–2011)

Sources: IEA (2010) U.S. Department of Energy (2010) for 2010 to 2011 estimates.

spent by the private sector. However, the private-sector share of the total has fallen over the last decade.

Estimates of private-sector spending further suggest that energy companies, at least in the United States, invest a far smaller share of sales in R&D than do high-technology industries such as the pharmaceutical, aircraft, or office equipment/computing industries.[6] Given the scale of the innovation challenge presented by current energy-related public policy concerns—particularly with respect to climate change—this observation prompts further questions: how can government stimulate additional private-sector investment in energy R&D? More specifically, what combination of "market-shaping" policies—including direct spending and incentives, as well as policies related to intellectual property, pricing and taxes, competition, technology mandates, and environmental standards and regulation—would most effectively accelerate the process of innovation and the introduction of innovative technologies to the marketplace? What is the overall level of private-sector R&D investment that could be brought to bear on the climate

6. This is notwithstanding the fact that many companies that provide energy-using goods and services—examples might include manufacturers of automobiles and electronic equipment—make substantial investments in R&D. In fact, some of these companies have very large R&D budgets (Newell 2010; U.K. Department for Innovation, Universities and Skills 2007). However, it is often difficult to discern what portion of the R&D budgets of major corporations goes to innovations that specifically affect the energy use characteristics of their product offerings.

technology challenge, and how does that level depend on the specific policy context in which companies make investment decisions (Newell 2010)?

Economists have investigated this process of induced innovation for many years in the context of a broad set of industries, and more-recent evidence supports the inducement mechanism specifically in the context of environmental and energy technology innovation in response to increases in cost of energy and environmental emissions (for surveys, see Jaffe, Newell, and Stavins 2003; Popp, Newell, and Jaffe 2010). Studies have, for example, looked at these questions using past examples of changes in regulatory or market conditions for energy technologies. The basic starting premise is that policies to address negative environmental externalities (such as standards or taxes) raise operating costs and create incentives for innovation. Indeed, a number of studies (e.g., Lanjouw and Mody 1996; Hascic, Johnstone, and Michel 2008; Popp 2006a) find that environmental regulations that impose emission reduction costs lead to increased private expenditures on abatement technologies and increased innovation (as measured by patents issued). Energy-related patenting activity also increases when energy prices rise, suggesting that policies that increase the cost of using fossil fuels can be expected to stimulate new research quickly (Popp 2002).

Other research suggests that changing regulatory conditions or simple uncertainty about future conditions tend to have a dampening effect on private-sector investment in new technologies. An analysis of data from the U.S. electric industry by Sanyal and Cohen (2009) suggests that R&D efforts by electric utility companies declined precipitously during the decade from 1990 to 2000, in large part because of the advent of electric industry restructuring. This created uncertainty about future regulatory and market conditions, which tended to discourage longer-term investments, including investments in R&D. Once restructuring legislation was adopted, exposure to competition tended to depress R&D investment even further. Sanyal and Cohen conclude that a sharp reduction in utility R&D expenditures is likely a permanent consequence of efforts to restructure the industry in the 1990s.

1.4   U.S. Government Investment in Energy RD&D

U.S. Department of Energy energy research has gone through several transitions over the last three decades, both in terms of its relative focus on precommercial basic research versus technology demonstration and in terms of the emphasis placed on different technology areas (e.g., nuclear power, fossil fuels, energy efficiency, and renewables). During the Nixon administration in the early 1970s, the primary goal was energy independence. This goal quickly proved impractical, but U.S. policy—especially after the 1973 OPEC oil embargo—continued to stress the development of alternative liquid fuels until well into the 1980s. The emphasis on finding domestic alternatives to imported oil culminated in the creation of the Synthetic Fuels Corporation, which became emblematic of the large, expensive demonstration projects undertaken during this era.

The Synthetic Fuels Corporation (SFC) was established in 1980 as an independent, wholly federally owned corporation to help create a domestic synthetic fuel industry as an alternative to importing crude oil. Under political pressure to backstop international oil prices, the SFC established a production target of 500,000 barrels per day. It had a seven-member board of directors, one of whom was a full-time chairman, and all of whom were appointed by the president and confirmed by the Senate. The SFC had the authority to provide financial assistance through purchase agreements, price guarantees, loan guarantees, loans, and joint ventures for project modules. After predicting oil prices of $80 to $100 per barrel and a synfuel price of $60 per barrel, the SFC was crippled when oil prices plummeted to below $20 per barrel. It was eventually canceled in 1986 after several billion dollars in expenditures. Many experts have criticized the SFC as an example of a failed involvement of government in large-scale commercial demonstration, an area thought better left to the private sphere (Cohen and Noll 1991).

Under the Reagan administration, national energy policy and federal research were dramatically reoriented, with a new stress on long-term, precompetitive R&D and lower overall budgets. By the late 1980s and early 1990s, DOE spending had dropped to less than half the peak levels of a decade earlier, and congressional appropriations were beginning to emphasize environmental goals, with large expenditures for the Clean Coal Technology Demonstration Program. The shift away from a focus on energy independence and resource depletion to a greater emphasis on environmental goals, energy efficiency and renewable energy, public-private partnerships, and cost sharing continued over the course of the Clinton administration in the 1990s. Meanwhile, federal support for basic energy research continued to receive the most consistent levels of funding, including in recent years.

Attempts to analyze the success or cost-effectiveness of past federal research relating to energy and the environment have come to mixed conclusions. Cohen and Noll (1991) documented the waste associated with the breeder reactor and synthetic fuel programs in the 1970s (noted in the preceding), but Pegram (1991) concluded that the photovoltaics research program undertaken during the same time frame had significant benefits. More recently, the U.S. National Research Council (NRC) conducted a comprehensive overview of energy efficiency and fossil energy research at the DOE during 1978 to 2000 (NRC 2001). Using both estimates of overall return and case studies, the NRC concluded that there were only a handful of programs that proved highly valuable. Returns on these programs, however, were such that their estimated benefits—including substantial direct economic benefits as well as external benefits such as pollution mitigation and knowledge creation—justified the overall portfolio investment.

Specifically, the NRC found that R&D investments in three types of energy efficiency technologies—advanced refrigerator and freezer compressors, electronic ballasts for fluorescent lamps, and low-emissivity glass—delivered cumulative estimated cost savings on the order of $30 billion when coupled with efficiency standards mandating their deployment. This amount compares to an estimated DOE and private-sector investment in these technologies of only $12 million. By contrast, DOE investments in fossil energy R&D were far less successful. The NRC concluded that cumulative economic savings from these programs only barely exceeded costs (which totaled nearly $11 billion over the period 1986 to 2000), and most of those savings came from improved technologies for extracting oil and gas, not from efforts to develop alternative fossil energy supplies. For the period 1975 to 1985, which included the synfuels era, the DOE invested roughly $6 billion in fossil energy programs that yielded—according to the NRC estimates—about $3.4 billion in benefits.

Although some projects can be expected to fail in any R&D program, the DOE's approach to fossil fuel R&D prior to 1985, with its focus on a narrow set of very expensive projects, did not pay off.[7] Moreover, funding for some programs continued long after it was known that they were ineffective or unlikely to succeed. In some cases, this was for political reasons (Congress continued to appropriate funds for some programs even after the DOE recommended they be cancelled); to some extent, this occurred because neither the DOE, nor the outside agencies charged with evaluating the DOE, applied a consistent, comprehensive, and objective methodology for assessing the costs and benefits of different programs.

U.S. government-sponsored energy R&D programs are commonly thought to have improved substantially since the 1970s and early 1980s, both in terms of the way they are managed and in terms of the objectives they target. To address problems of waste, the DOE launched a series of reforms in the 1990s that were intended to strengthen its contracting and project management practices, hold contractors more accountable for their performance, and demonstrate progress in achieving the agency's missions (Norberg-Bohm 2000; Wells 2001). The improvement in the DOE's more recent track record—particularly with respect to its fossil energy programs—may also be attributed to the shift that occurred in the essential nature of the agency's R&D portfolio during the 1980s. According to the NRC study:

> The fossil energy programs of the 1978 to 1986 period, which was dominated by an atmosphere of crisis following the 1973 oil embargo, empha-

7. As the authors of the NRC report point out, an R&D strategy that never produced any failures would not be desirable either; rather, it would indicate an overly conservative approach to the selection of research priorities that almost surely would result in missed opportunities. Rather than striving to minimize risk and avoid failure, the NRC recommends a portfolio approach that emphasizes diversity, goal-setting, objective assessment, and performance tracking.

sized a high-risk strategy for circumventing commercial-scale demonstrations by going directly from bench-scale to large-scale demonstrations to make synthetic fuels from coal and shale oil and to produce oil using enhanced oil recovery techniques. In the second period, however, the fossil energy R&D program was systematic and involved a more diverse portfolio and greater emphasis on increasing the efficiency of electric power generation using natural gas, on reducing the environmental impact when burning coal, and on advanced oil and gas exploration and production. (NRC 2001, 63)

Despite this shift, interest in large-scale, government-sponsored demonstration projects has continued. A recent example is the FutureGen Initiative, which was launched in 2003 as a public-private effort to demonstrate a near-zero-emission, 275 megawatt (MW) coal-fired power plant for producing hydrogen and electricity with carbon capture and storage. FutureGen has already had a turbulent history: By the end of 2007, a consortium of thirteen power producers and electric utilities from around the world had agreed to participate, and a project site had been selected in Illinois. In January 2008, the DOE—citing cost concerns—abruptly cancelled funding for the project. In June 2009, the Obama administration announced its intent to reinstate federal funding for FutureGen; shortly thereafter, however, two large U.S. utility companies—American Electric Power and Southern Company—withdrew from the project (in all, four participants have withdrawn, leaving a total of nine companies in the FutureGen Alliance). In addition, a number of controversies have arisen in connection with the project design, including the choice of a project site, the size of the federal cost-share, the fraction of carbon dioxide emissions to be captured and stored, and project cost.

A small number of papers have also attempted to evaluate the success of government efforts to accelerate the "transfer" of knowledge from basic to applied research (a step that can be seen as bridging the processes of invention and innovation). Such efforts typically combine basic and applied research and are often implemented through government-industry partnerships (National Science Board 2006). The United States passed several policies in the 1980s specifically designed to improve transfer from the more basic research done at government and university laboratories to the applied research done by industry to create marketable products.

Jaffe and Lerner (2001) studied the effectiveness of DOE-funded research and development centers in this regard, supplementing a detailed analysis of patents assigned either directly to the laboratories or to private contractors who collaborated on research at the labs with case studies of two DOE laboratories where technology transfer efforts increased in the 1980s and 1990s. They find that both the number of patents obtained and the number of citations received per patent increased at DOE laboratories since the policy shifts of the 1980s. That the number of citations also increased after the 1980 policy changes contrasts with the findings of researchers who have studied academic patenting, where patent activity increases over time, but the quality of patents appears to decline. Jaffe and Lerner also find that the type of research performed at a laboratory affects technology transfer. Transfer is slower when more basic research is performed or when the research has national security implications. Interestingly, the national laboratories with greater contractor turnover appeared to be more successful at commercializing new technologies.

Popp (2006b) examined citations made to patents in eleven energy technology categories, such as wind and solar energy. He finds that energy patents spawned by government R&D are cited more frequently than other energy patents. This is consistent with the notion that these patents are more basic. More important, after passage of the technology transfer acts in the early 1980s, the privately held patents that are cited most frequently are those that themselves cite government patents. This suggests that publicly sponsored research continues to provide benefits even after the results of that research are transferred to private industry.

1.5   Conclusion

Even a cursory review of the history of energy technology suggests tremendous potential for innovation, both in the technologies available for energy production and in the technologies for energy use. Where a market exists or emerges for technological improvements, innovation has produced significant gains. Thus, for example, advances in the tools and techniques available for extracting energy resources like oil and natural gas have made it possible for accessible reserves to keep pace with rising demand for these fuels over time. However, the most pressing energy challenges that now confront humanity involve environmental and other societal externalities for which there has historically been little or no market.

Among those challenges is climate change, which has emerged—alongside continuing concerns about energy supply security—as one of the central issues motivating most current discussions about energy technology innovation. The remainder of this book explores patterns of technological innovation in other industries to see what lessons might be applicable in the energy context and, more specifically, to understand what roles government and the private sector might play in accelerating the process of innovation. Both theory and empirical evidence suggest that the public role has at least two dimensions: (a) creating a market for technological improvements through policy intervention (environmental regulation provides a classic example) and (b) investing directly in innovation, for example, through support for R&D, which tends to be underprovided if left to the private sector alone. The case for public investment in R&D is based on knowledge spillovers and other societal benefits; it is the subject of a well-established economic literature.

In the first role—eliciting technological innovation through policies and regulations—governments in the developed world have been, on the whole, quite effective. Very substantial improvements in efficiency and environmental performance have been achieved across a wide array of energy production and end-use technologies in response to various standards and other requirements. A number of studies over the past several years have also evaluated the performance of federal energy R&D programs. Although these R&D programs have produced some notable failures and although their performance has varied widely, these evaluations support the finding that federal energy R&D investments have yielded, on the whole, substantial direct economic benefits as well as external benefits such as pollution mitigation and knowledge creation. However, as the NRC concluded in its study of DOE's fossil fuel and efficiency R&D programs, "forced" government introduction of not-yet-economic new technologies has not been successful (also see Fri 2003).

In addition, suggestions for strengthening the organization, management, and priorities of federal energy R&D efforts emerge from every recent major study of these activities (Newell 2008; Ogden, Podesta, and Deutch 2008; Chow and Newell 2004; National Commission on Energy Policy 2004). Headway has been made at the DOE along several of these lines, and a number of provisions in the Energy Policy Act of 2005 codify recent trends in research management, including nonfederal cost-sharing for projects, increased merit review and competitive award of proposals, external technical review of departmental programs, and improved coordination and management of programs. Interest has also increased in further cultivation of partnerships linking firms, national laboratories, and universities. Particularly in the context of increasing the transfer of knowledge to technology application, experts have highlighted the importance of improving processes for communication, coordination, and collaboration within the DOE among the basic research programs in the Office of Science and the applied energy research "stovepipes" within the DOE program offices (fossil fuel, nuclear, renewables, end-use efficiency, electricity reliability).

The lessons from past private and public innovation efforts suggest that a well-targeted set of climate policies, including those targeted directly at science and innovation, could help lower the overall costs of climate change mitigation. It is important to stress, however, that poorly designed technology policy could raise rather than lower the societal costs of climate mitigation. To avoid this, policymakers may want to examine the idea of creating substantial incentives in the form of a market-based price on GHG emissions. Furthermore, directed government technology support has been shown to be most effective when it emphasized areas least likely to be undertaken by a private sector. As discussed, this would tend to emphasize use-inspired basic research that advances science in areas critical to climate mitigation and other energy goals. In addition to generating new knowledge and useful tools, such funding also serves the critical function of training the next generation of scientists and engineers for future work in the private sector, at universities, and in other research institutions. As the largest single supporter of U.S. basic research in the physical sciences—accounting for 40 percent of federal outlays in this area—the DOE Office of Science has an important role in this process.

Innovation policy has been most efficient in the energy arena when it has complemented rather than attempting to directly substitute for market demand. Nonetheless, R&D without market demand for the results is like pushing on a rope and has resulted in little impact. The scale of the climate technology problem and our other energy challenges suggests a solution that maximizes the impact of the scarce resources available for addressing these and other critical societal goals. Evidence indicates that an emissions price plus RD&D approach could provide the basic framework for such a solution.

References

Ausubel, Jesse, and Cesare Marchetti. 1996. Elektron: Electrical systems in retrospect and prospect. *Daedalus: Journal of the American Academy of Arts and Sciences* Summer:139–69.

Cambridge Energy Research Associates (CERA). 2006. *Why the peak oil theory falls down: Myths, legends, and the future of oil resources.* Cambridge Energy Research Associates Report. Cambridge, MA: CERA.

Carlson, Curtis, Dallas Burtraw, Maureen Cropper, and Karen L. Palmer. 2000. Sulfur dioxide control by electric utilities: What are the gains from trade? *Journal of Political Economy* 108 (6): 1292–1326.

Chow, Jeffrey, and Richard G. Newell. 2004. A retrospective review of the performance of energy R&D. Resources for the Future, Discussion Paper. Washington, DC: Resources for the Future.

Cohen, Linda R., and Roger G. Noll. 1991. *The technology pork barrel.* Washington, DC: Brookings Institution.

Deutch, John M., Charles W. Forsberg, Andrew C. Kadak, Mujif S. Kazimi, Ernest J. Moniz, John E. Parsons, Yangbo Du, and Laura Pierpoint. 2009. Update of the 2003 future of nuclear power. Cambridge, MA: MIT Energy Initiative.

Economist. 2008. Wind of change. December 4. http://www.economist.com/display Story.cfm?story_id=12673331.

Fri, Robert W. 2003. The role of knowledge: Technological innovation in the energy system. *Energy Journal* 24 (4): 51–74.

Hascic, Ivan, Nick Johnstone, and Christian Michel. 2008. Environmental policy stringency and technological innovation: Evidence from patent counts. Paper presented at the European Association of Environmental and Resource Economists 16th annual conference, Gothenburg, Sweden.

InterAcademy Council. 2007. Lighting the way: Toward a sustainable energy future. Amsterdam: InterAcademy Council Secretariat. http://www.interacademycouncil .net/?id=12198.

Jaffe, Adam B., and Josh Lerner. 2001. Reinventing public R&D: Patent policy and the commercialization of national laboratory technologies. *RAND Journal of Economics* 32 (1): 167–98.

Jaffe, Adam B., Richard G. Newell, and Robert N. Stavins. 2003. Technological change and the environment. In *Handbook of environmental economics*. Vol. 1, ed. Karl-Goran Mäler and Jeffrey Vincent, 461–516. Handbooks in Economics, ed. K. J. Arrow and M. D. Intriligator. Amsterdam: North-Holland/Elsevier.

Lanjouw, Jean Olson, and Ashoka Mody. 1996. Innovation and the international diffusion of environmentally responsive technology. *Research Policy* 25 (4): 549–71.

Mouawad, Jad. 2007. Oil innovations pump new life into old wells. *New York Times,* March 5. http://www.nytimes.com/2007/03/05/business/05oil1.html?_r=2&scp =4&sq=chevron+bakersfield&st=nyt.

National Commission on Energy Policy. 2004. *Ending the energy stalemate: A bipartisan strategy to meet America's energy challenges.* National Commission on Energy Policy Report. Washington, DC: National Commission on Energy Policy.

National Renewable Energy Lab (NREL). 2008. NREL solar cell sets world efficiency record at 40.8 percent. News release, August 13. Golden, CO: National Renewable Energy Lab, U.S. Department of Energy, Office of Energy Efficiency and Renewable Energy.

National Research Council (NRC). 2001. *Energy research at DOE, was it worth it? Energy efficiency and fossil energy research 1978 to 2000.* Washington, DC: National Academies Press.

National Science Board. 2006. Research and development: Funds and technology linkages. In *Science and engineering indicators 2006.* Arlington, VA: National Science Foundation.

Neij, Lena. 1999. Cost dynamics of wind power. *Energy* 24 (5): 375–89.

———. 2005. International learning with wind power. *Energy and Environment* 15 (2): 175–86.

Newell, Richard G. 2008. A U.S. innovation strategy for climate change mitigation. Hamilton Project Discussion Paper no. 2008-15. Washington, DC: Brookings Institution.

———. 2010. The role of markets and policies in delivering innovation for climate change mitigation. *Oxford Review of Economic Policy* 26 (2): 253–69.

Norberg-Bohm, Vicki, ed. 2002. The role of government in energy technology innovation: Insights for government policy in the energy sector. Energy Technology Innovation Project, Belfer Center for Science and International Affairs, John F. Kennedy School of Government, Harvard University, Working Paper no. 2002-14. Cambridge, MA: Harvard University Press.

Ogden, Peter, John Podesta, and John Deutch. 2008. A new strategy to spur energy innovation. *Issues in Science and Technology* Winter: http://www.issues.org/24.2/ ogden.html.

Pegram, William M. 1991. The photovoltaics commercialization program. In *The technology pork barrel,* ed. Linda R. Cohen and Roger G. Noll, 321–63. Washington, DC: Brookings Institution.

Popp, David. 2001. The effect of new technology on energy consumption. *Resource and Energy Economics* 23:215–39.

———. 2002. Induced innovation and energy prices. *American Economic Review* 92 (1): 160–80.

———. 2003. Pollution control innovations and the Clean Air Act of 1990. *Journal of Policy Analysis and Management* 22 (4): 641–60.

———. 2006a. International innovation and diffusion of air pollution control technologies: The effects of NOX and SO2 regulation in the U.S., Japan, and Germany. *Journal of Environmental Economics and Management* 51 (1): 46–71.

———. 2006b. They don't invent them like they used to: An examination of energy patent citations over time. *Economics of Innovation and New Technology* 15 (8): 753–76.

Popp, David, Richard G. Newell, and Adam B. Jaffe. 2010. Energy, the environment, and technological change. In *Handbook of the economics of innovation.* Vol. 2, ed. Bronwyn H. Hall and Nathan Rosenberg, 873–937. Amsterdam: Elsevier B. V.

Renewable Energy Policy Network for the 21st Century (REN21). 2009. Renewables global status report. 2009 update. Paris: REN21 Secretariat. http://www.world watch.org/files/pdf/RE_GSR_2009.pdf.

Sanyal, Paroma, and Linda R. Cohen. 2009. Powering progress: Restructuring, competition, and R&D in the U.S. electric utility industry. *Energy Journal* 30 (2): 41–80.

Stavins, Robert N. 1998. What can we learn from the grand policy experiment? Lessons from SO2 allowance trading. *Journal of Economic Perspectives* 12 (3): 69–88.

U.K. Department for Innovation, Universities and Skills. 2007. The 2007 R&D scorecard. London: U.K. Department for Innovation, Universities and Skills.

United Nations Development Programme (UNDP). 2000. *World energy assessment: Energy and the challenge of sustainability,* ed. Jose Goldemberg. New York: United Nations.

U.S. Department of Energy (DOE). 2009. *2008 renewable energy data book.* Washington, DC: Office of Energy Efficiency and Renewable Energy. http://www1.eere .energy.gov/maps_data/pdfs/eere_databook.pdf.

———. 2010. FY2011 congressional budget request: Budget highlights. Washington, DC: U.S. Department of Energy.

U.S. Energy Information Administration (EIA). 2006. Iron and steel manufacturing energy intensities, 1998 and 2002. Washington, DC: U.S. Energy Information Administration. http://www.eia.doe.gov/emeu/efficiency/iron_steel_9802/steel _9802_data.html.

———. 2009a. *Annual energy review.* Washington, DC: Energy Information Administration.

———. 2009b. *International energy outlook 2009.* Washington, DC: Energy Information Administration.

———. 2009c. International energy statistics. Washington, DC: Energy Information Administration. http://tonto.eia.doe.gov/cfapps/ipdbproject/IEDIndex3.cfm?tid =2&pid=2&aid=12.

———. 2009d. *Monthly energy review.* Washington, DC: Energy Information Administration, July.

———. 2010. *International energy outlook 2010.* Washington, DC: Energy Information Administration.

U.S. Environmental Protection Agency (EPA). 2009. *Acid rain and related programs: 2007 progress report.* Report no. EPA-430-K-08-010, January. Washington, DC:

Environmental Protection Agency. http://www.epa.gov/airmarkt/progress/docs/2007ARPReport.pdf.

U.S. Geological Survey (USGS) World Energy Assessment Team. 2000. U.S. Geological Survey world petroleum assessment 2000—Description and results. U.S. Geological Survey Digital Data Series no. DDS-60. Washington, DC: U.S. Geological Survey. http://energy.cr.usgs.gov/WEReport.pdf.

Watanabe, Chihiro, Kouji Wakabayashi, and Toshinori Miyazawa. 2000. Industrial dynamism and the creation of a "virtuous cycle" between R&D, market growth and price reduction: The case of photovoltaic power generation (PV) development in Japan. *Technovation* 20 (6): 299–312.

Wells, Jim. 2001. Fossil fuel R&D: Lessons learned in the Clean Coal Technology Program. Testimony before the Subcommittee on Energy, Committee on Science, House of Representatives. GAO-01-854T. Washington, DC: General Accounting Office.

World Nuclear Association. 2005. Outline history of nuclear energy. London: World Nuclear Association. http://www.world-nuclear.org/info/inf54.html.

2

Agricultural Innovation

Tiffany Shih and Brian Wright

2.1 Introduction

Agricultural production is a highly decentralized and geographically dispersed activity, dependent upon a wide variety of technologies applied to a heterogeneous natural resource base that is changing over time. Its principal products are necessary inputs into consumer goods essential for life, and continuous availability at acceptable quality and price is vital.

The history of agricultural innovation is relevant to plans for accelerating energy innovation because energy production and use share many of the preceding characteristics. Many innovative energy technologies are becoming more similar to agriculture as they revert from depletable to renewable resource bases. Indeed, the principal currently commercialized biofuels technologies involve agricultural production. Like agriculture, energy production critically relates to greenhouse gas production. It is widely agreed that climate change will have a greater impact on agricultural production than on any other sector, while in the energy sector, its effect on investment is already significant.

The record of achievement in agricultural innovation over the past century is impressive. Increases in agricultural productivity have fueled rates of increase in food supply that outpaced the joint effects of growth in personal consumption and population, with only modest recruitment of new cropland (Pardey and Beintema 2001). Better nutrition has, in turn, transformed

Tiffany Shih is a PhD student in the Department of Agricultural and Resource Economics at the University of California, Berkeley. Brian Wright is professor and chair of the Department of Agricultural and Resource Economics at the University of California, Berkeley.

Preliminary work on which this chapter is based was supported by the Energy Biosciences Institute and the National Institutes of Health.

life expectancies, labor productivity, and the rate of population growth. Agricultural research activities have spread around the globe with marginal social rates of return so high that they strain credulity. Patterns of participation and technology exchange demonstrate high interdependence both between countries and along the public-private spectrum (Wright et al. 2007, table 1).

Agriculture has a long history of productive public research. Evenson's (2001) survey of over forty studies between 1915 and 1999 gives a marginal real social rate of return to U.S. public agricultural research investments between 45 to 65 percent.[1] Alston et al. (2010) report marginal internal rates of return on state agricultural research to be, on average, at least 19 percent for private returns and at least 23 percent for public returns.[2] The development of an effective system for public investment in research and knowledge dissemination critically contributed to the observed pace of agricultural advancement. In addition, sharing of knowledge and innovation between farmers, input suppliers, and researchers, both within agriculture and beyond, has played an important role. For example, just as applied research in electronics, communications, and nuclear energy have drawn from basic research funded by the U.S. Department of Defense, recent innovations in agricultural biotechnology owe much to projects on human health funded by the National Institutes of Health.

Compared to other sectors, agricultural research investments are more geographically dispersed, both within the United Stated and globally. Indeed, public investments in agricultural research from developing countries have recently overtaken public agricultural investment in developed countries as a whole (Wright et al. 2007). The fundamental influence from the natural features of the growth environment means that adaptive research is often needed to apply agricultural biotechnological innovations in a given local area. The U.S. institutional structure supporting agricultural research reflects this reality, employing state- and local-level research institutions and experiment stations to meet the needs of local farmers. Because the primary benefit of innovation in agriculture is lower food prices, countries with large populations can internalize a larger share of the gains from more basic research that has strong externalities across neighboring environments. Thus, large countries account for a dominant portion of public research investments.

Beginning in the United States in 1980, advances in biotechnology, combined with the extension of strong intellectual property (IP) protection to

1. However, in a meta-analysis of nearly 300 relevant studies (including those derived from U.S. data or foreign data), Alston et al. (2000a) note that the estimates for annual rates of return range from –7.4 to 5,645 percent. Few studies actually fall into the 45 to 65 percent category. The authors find significant variation in the estimates derives from the different rate of return measures used, potential analyst biases and methodologies used, and the type of research evaluated. Alston et al. (2000b) instead estimate an 80 percent overall annual rate of return to research.

2. Alston et al. (2010, table 11-4, 368, and table 11-5, 369).

agriculture, have elicited such a great private investment response that it now exceeds public investment. The proliferstions of IP, however, has not been without consequences. There is mounting evidence of a negative effect on freedom to operate in public-sector projects directed toward production of plant varieties and other technologies for use by farmers. The problem arises from fragmented IP claims to technology inputs and is especially prominent in the agricultural biotechnology field. The private sector responded to costly sharing of IP with a series of industry purchases, leading to market concentration and the current dominance of the agricultural industry by the firm Monsanto.

Public interventions with other objectives also affect innovation. Public support for the incomes of farmers and other agricultural investors moderates their opposition to policies reducing commodity prices, revenues, and, more fundamentally, asset values. Government regulations regarding food production and distribution, employment, environmental protection, and intellectual property rights (IPRs) also affect the level and distribution of private research and development and technology investments on and off the farm, as well as the market structure of input industries.

Like the energy sector, the agricultural sector is facing continued challenges posed by a growing global population with changing demands on the amounts and kinds of resources being used and by evolving concerns regarding global resources and environmental constraints. Awareness about global interdependence of the world's food supplies has extended to concerns regarding environmental externalities such as global warming, motivating international efforts to harmonize regulations and share technologies. Recently, concern about food security has engendered a flurry of private investment and strategic activity.

In this chapter, we provide an overview of innovation policies for agriculture with the purpose of highlighting aspects of interest to the energy sector. We proceed as follows. Section 2.2 provides a summary of global agricultural research investments with particular attention to public versus private investments and their distribution between developed and less-developed countries. Section 2.3 provides a brief history of U.S. public investment in agriculture. We next discuss the main intellectual property mechanisms relevant to agriculture and their effects on research and innovation markets, followed by a short discussion of the extent and efficacy of public-private collaborations in section 2.5. Section 2.6 considers government regulation and its dual responsibilities of ensuring public safety and promoting technological advances. We then describe the factors influencing technological adoption in agriculture, followed by a brief conclusion.

2.2   Agricultural Research Investments

According to the latest available figures, public and nonprofit funding accounts for about two-thirds of global agricultural research expenditures, while private investments account for the other third (Wright et al. 2007). These aggregate figures mask the fact that the *types* and *scope* of research performed by the two sectors are neither perfectly substitutable nor independent. In general, private investments are far more concentrated by crop and technology than public investments and tend to rely on the latter for the basic science inputs in order to produce applied technology (Alston, Pardey, and Roseboom 1998).

2.2.1 Research Investment Trends

Public spending on agricultural research experienced overall growth in the past few decades, from about US$ 15.2 billion (at year 2000 prices) in 1981 to US$ 23 billion in 2000 (table 2.1). During the 1990s, public agricultural research expenditures by developing countries as a group exceeded those by developed countries for the first time. In no other sector does research expenditure by developing countries have comparable prominence.

China, India, Brazil, South Africa, and Thailand now undertake over half the investment in less-developed countries. However, research per capita or relative to agricultural output is far less concentrated. Similarly, public

Table 2.1   **Estimated global public and private agricultural research and development (R&D) investments, 2000**

	Agricultural R&D spending (international 2000$ million)		Shares in global total (%)	
	1981	2000	1981	2000
Public				
Asia and Pacific (28)	3,047	7,523	20.0	32.7
Latin America and Caribbean (27)	1,897	2,454	12.5	10.8
Sub-Saharan Africa (44)	1,196	1,461	7.9	6.3
West Asia and North Africa (18)	764	1,382	5.0	6.0
Subtotal, less-developed countries (117)	6,904	12,819	45.4	55.8
United States	2,533	3,828	16.7	16.6
Subtotal, high-income countries (22)	8,293	10,191	54.6	44.2
Total (139)	15,197	23,010	100.0	100.0
Private				
Developing		869		6.5
High income		12,577		93.5
Total		13,446		100.0
Public and private				
Developing		13,688		37.5
High Income		22,767		62.5
Total		36,456		100.0

Source: Pardey et al. (2006).

spending by developed countries is concentrated on a small set of countries: the United States, Japan, France, and Germany. Developed countries spent about as much on public agricultural research (US$ 12,577 million) in 2000 as developing countries spent on public research (US$ 12,819 million). Private spending in developing countries was only US$ 819 million, as shown in table 2.1. (Wright et al. 2007; Wright and Pardey 2006)

As in energy, public and private research spending tends to increase in response to periods of high prices and go slack when low prices have restored a false sense of security. Public research spending, strong after the food price crises of the 1970s and early 1980s, declined during the 1990s in most areas. In developed countries, annual growth rates of 2.2 percent during the 1980s fell to 0.2 percent per year from 1991 to 1996. In Africa, growth in agricultural research spending ground to a halt in the 1990s, with some revival more recently. Spending in China and Latin America, on the other hand, grew in the early 1990s after stagnating in the 1980s. China has been particularly focused on the agricultural biotechnology field, increasing spending from $17 million in 1986 to nearly $200 million by 2003 (Huang et al. 2002; Huang et al. 2005).

Other measures of research investments reveal sharp contrasts between wealthy and poor nations. Both developed and less-developed countries increased spending on public agricultural research and development (R&D) per dollar agricultural output in the past few decades. In developed countries, spending on public agricultural R&D per-dollar agricultural output increased to $2.64 per $100 agricultural output in 1995 from $1.53 per $100 of output in 1975. In the developing world over this interval, growth in research intensities also increased, on average, but the level was much lower and varied between countries. Growth was constant in China, increasing in other parts of Asia and in Latin America, but decreasing significantly in Africa.

While the rates of research expenditures are informative about recent policies, researchers have found the accumulated stock of research capital to be a more relevant determinant of research capabilities. Pardey and Beintema (2001) calculate that the agricultural research resource stock, as a proportion of the value of agricultural output, is at least twelve times larger in the United States than in Africa, given reasonable rates of interest and depreciation.[3]

2.2.2   International Funding Organizations

International funding agencies have been established to direct more research resources toward more efficiently serving the needs of poorer

---

3. For further details, see Wright et al. (2007) and Wright and Pardey (2006) from which this section is largely drawn. See also Pardey, Alston, and Piggott (2006) for a discussion of agricultural research investments in less-developed countries.

nations. Most notably, 1971 saw the establishment of the Consultative Group on International Agricultural Research (CGIAR or CG), an international partnership between governments, private foundations, and civil society organizations that fund and influence the research of fifteen international agricultural research centers (IARCs). The roots of CGIAR are in the 1943 International Agriculture Program, a cooperative effort initiated by the U.S. and Mexican governments with significant support from the Rockefeller Foundation. The Mexican program became the International Center for the Improvement of Maize and Wheat (CIMMYT). The relatively simple funding and managerial structure proved superior to the less-focused agenda of the Food and Agriculture Organization (FAO) in producing major agricultural innovations. It developed high-yielding semidwarf wheat varieties that were more responsive to nitrogen fertilizer. Contemporaneous innovation in the fertilizer sector enabled production of cheaper nitrogen fertilizer from natural gas. The technology package offering cheaper nitrogen fertilizer, and the wheat varieties that could exploit it laid the basis for the Green Revolution in wheat in Mexico and Asia. Desire to broaden the scope of yield increases to other crops prompted the establishment of similarly structured research centers in Colombia, Nigeria, and the Philippines by 1971.

The need for a broader funding base to support the new centers led the International Agriculture Program to enlist the participation of a large range of other donors, including several national aid organizations, the United Nations FAO, the International Fund for Agricultural Development, the United Nations Development Program, and the World Bank, in the establishment of the CGIAR (CGIAR 2008). After establishment, the annual spending of the members of the CGIAR grew rapidly, from $1.3 million supporting four centers in 1965, to $141 million supporting eleven centers by 1980, and to $305 million supporting thirteen centers by 1990. Growth slowed in the 1990s so that spending per center during this decade declined. Between 1994 and 2002, and in response to critiques of the Green Revolution, complacency regarding food supplies, and the conflicting agendas and diverse interests of funding entities, total CGIAR funding dropped in real terms. Investments in germplasm enhancement research declined at 6.5 percent per year during this time, reflecting a shifted focus to policy and environmental objectives. This pattern was mirrored by similar declines in agricultural aid and research funding from the European Community, the World Bank, and the United States Agency for International Development (USAID) from the mid-1980s through the 1990s. More recently, in response to high agricultural prices and a renewed recognition of the importance of an adequate food supply, funding has begun to trend upward again. CGIAR funding was over $495 million in 2007, and World Bank agricultural lending increased from an annual average of $1.5 billion in 2002 to $4.6 billion between 2006–2008. (Wright et al. 2007; Lele 2003; Lele et al. 2003; CGIAR Independent Review Panel 2008; World Bank Group 2009).

2.3   A History of U.S. Public Agricultural Support

The U.S. made an early commitment to agricultural research in the form of agricultural institutional innovations adopted from Europe. Economists have emphasized the success of the land grant college system (National Research Council 1995; Huffman and Evenson 2006; Ruttan 1980; Ruttan 1982), noting its contribution to lowered food costs, rural development, and the prominence of U.S. agriculture globally (Adelaja 2003). A lesson embedded in the long history of U.S. public support for agriculture is the importance of a large buildup of research capital stock through sustained investment. In addition, the nationwide adoption of agricultural innovations was encouraged by a decentralized institutional system capable of adapting technology to local environmental conditions, as was incorporated in the land grant colleges and state agricultural experiment stations.

2.3.1   The Establishment of U.S. Agricultural Institutions

Spatial environmental variation forms the context in which technology and resources determine agricultural productivity. The expansion of arable land followed by mechanical innovations produced by farmers and blacksmiths drove early increases in U.S. agricultural output (Huffman and Evenson 2006; Sunding and Zilberman 2001). Major increases in yields *per acre* were not achieved until the advances in hybrid seed and agrichemical technology in the 1930s. Until that time, biological innovation focused on disseminating and adapting crops to unfamiliar frontier environments (Olmstead and Rhode 1993).

Human efforts to locate and distribute plant genetic material appropriate for given production environments or meeting particular consumer needs originated long before the groundbreaking discoveries of Mendel and Darwin, not to mention recent work in genetically engineered crops.[4] Heightened recognition of the economic value of plants in the context of the Industrial Revolution and their scientific documentation and classification encouraged the spread of botanic gardens across Europe in the eighteenth and nineteenth centuries. In particular, Britain's Royal Botanic Gardens at Kew excelled in the acquisition, development, and dissemination of economically important plants (Juma 1989). A physician's experimentation with urban plant cultivation in the polluted atmosphere of London during the Industrial Revolution led to the invention of the Wardian case or terrarium, an enclosed container that vastly increased the reliability of international transportation of live plants between the new and old worlds (Schoenermarck 1974).

In the United States, prominent figures such as George Washington,

---

4. Juma (1989) gives several examples of early plant-collecting expeditions and gardens spanning Ancient Egypt to Japan to colonial explorations of the Americas.

Benjamin Franklin, and Thomas Jefferson all recognized the benefits of acquiring diverse plant resources and endeavored to introduce improved plant varieties into the country. Jefferson himself once wrote, "The greatest service which can be rendered any country is to add a useful plant to its culture" and went so far as to smuggle rice from the Piedmont region of Italy into the United States, sewn into the lining of his coat pockets, when such a crime was punishable by death (Fowler 1994).[5] His enthusiasm for the importance of plant resources was shared by Henry Ellsworth, the first commissioner of the Patent Office and founder of what ultimately became the United States Department of Agriculture. Without congressional approval, Ellsworth distributed seeds and plant material from other lands in order to test and promote their benefits. The U.S. Patent Office thus became the main repository for plant genetic material in the country, relying on the U.S. Navy to import foreign seed and the U.S. Post Office to distribute seeds to farmers through the mail. Producing a number of documents on proven and potential economic benefits of plant resources, Ellsworth championed federal support for agriculture and the creation of an independent national agricultural research bureau. As a result, in 1839, Congress began to formally support seed collection, distribution, and research efforts by establishing the Agricultural Division of the Patent Office, which became the Department of Agriculture in 1862. (Harding 1940; Huffman and Evenson 2006).

This widespread recognition of plant resource benefits plus the dominance of the U.S. farmer population created a favorable political climate in support of the passage of the foundational 1862 Morrill (Land Grant Colleges) Act (7 U.S.C. § 301 et seq) and the 1887 Hatch Act (7 U.S.C. § 361a et seq). The Morrill Act allotted federal land to each state to support the development of a college focused on instruction in "agriculture and the mechanic arts." (7 U.S.C. § 304). Originally blocked by southern states where education was viewed as a threat to the cheap labor supply, the Morrill Act was passed after the South seceded. The Hatch Act provided additional federal lands to conduct and disseminate research in State Agricultural Experiment Stations (SAESs) associated with land grant colleges. In recognition of the importance of technology transfer mechanisms for realizing the benefits of research, the 1914 Smith-Lever Act established the Cooperative Extension Service to distribute knowledge relevant for the local adoption and application of innovations. These key acts balanced federal and state roles by combining federal financial support with state management for the administration and direction of research. The resulting structure provided an avenue to address local research needs while also exploiting interstate competition to motivate fruitful research. As early as 1888, states began to establish substations that addressed needs distributed at even finer geographic scales (Huffman and Evenson 2006; Ruttan 1982).

---

5. Thomas Jefferson, *The Works of Thomas Jefferson,* vol. 8, federal ed. (New York: G.P. Putnam's Sons, 1904–1905). http://oll.libertyfund.org/title/805.

The case of hybrid corn exemplifies the advantage of regionally focused agricultural research that benefited both local farmers and consumers generally. This innovation, which originated as a by-product of basic research in genetics conducted at Harvard decades earlier (Troyer 2009), required additional decades of adaptive research after its initial adoption in the heart of the midwest corn belt to spread across the states in which it was ultimately established (Griliches 1957).

The establishment of the SAES system in the United States borrowed heavily from European developments that firmly established the central role of universities and scientists in agricultural development. Justus von Leibig, a German chemist who founded the first modern chemistry laboratory, became one of the forefathers of agricultural science with his 1840 publication *Organic Chemistry in Its Relation to Agriculture and Physiology* (Brock 1997). During this time, agricultural research institutions in the states that eventually formed Germany demonstrated the potential power of a group of experts working on a focused field, highlighted the importance of consistent funding, provided valuable experience navigating the link between science and practice, and demonstrated the merits of interinstitutional competition. By the time of Liebig's death in 1873, the newly united Germany had twenty-five agricultural research stations. The German development of successful university-based agricultural chemistry research laboratories and experiment stations became the model followed throughout the United States and Europe, where numerous agricultural experiment stations were also established during the second half of the nineteenth century. In particular, Rothamsted Agricultural Experiment Station in England, currently the oldest continuously operating agricultural experiment station, was founded by a fertilizer manufacturer in 1843 (Huffman and Evenson 2006; Finlay 1988).

The first U.S. stations continued the heavy emphasis on agricultural chemistry established in Germany, and Samuel W. Johnson, the first director of a U.S. agricultural experiment station, was trained by a founder of the German system. By the time the Hatch Act was passed, fifteen primarily state-funded experiment stations were already in operation.

Major benefits of public research for agricultural productivity in the United States began to accrue only after the research establishment had accumulated a substantial stock of knowledge. Evenson (1980) found that during 1870 to 1925, agricultural productivity was strongly correlated with the total real public-agricultural research spending over the preceding eighteen years. Early advances in U.S. agriculture were largely borrowed from progress made in Europe. It took several decades of development and learning before the U.S. land grant/SAES system had acquired the scientific capacity and research base necessary to become an efficient system of innovation (Huffman and Evenson 2006). Subsequent research has confirmed that stable agricultural funding promotes persistent gains in agricultural productivity (White and Havlicek 1982).

2.3.2   Private Interests and the Allocation of Public Funding

Because private research focuses mainly on commercializable technologies with appropriable benefits, the onus is on the public sector to produce basic science and undertake research that may be high in risk, have long lag times, or create unpredictable and nonexcludable benefits (Alston, Pardey, and Roseboom 1998; Stokes 1997; Huffman and Evenson 2006; Just and Huffman 1992). High rates of return to investments in different types of agricultural technologies persist, implying that those investments have overall made excellent use of public funds, but also that the level of funding has been inadequate (Judd, Boyce, and Evenson 1986; Huffman and Evenson 2006). In addition, in cases where public support for research has declined and there are private incentives or IPRs for innovation, some observers have concluded that public research funds have been increasingly directed toward the development of private goods (for example, Knudson and Pray 1991). Economists have warned that heightened private influence over public research agendas may further erode public support for research, thus damaging the system by which basic science and public goods research is produced (Just and Huffman 1992). However, Foltz, Barham, and Kim (2007) found evidence of economies of scale and scope in life science research production of patents and journal articles, particularly in land grant universities, implying synergies may exist in university production of private and public goods.

A second concern related to public funding has been discussed in a number of empirical studies that suggest interregional externalities in the United States significantly affect state research investment levels (Guttman 1978; Huffman and Miranowski, 1981; Rose-Ackerman and Evenson 1985). For example, citing unpublished work, Alston (2002) finds that averaging across U.S. states, over half the measured within-state productivity gains may be derived from the benefits of research investments made elsewhere. Such spillovers, both within and among nations, may contribute to underinvestments in research. In addition, studies by Hayami and Ruttan (1971) and Binswanger and Ruttan (1978) suggest that nationally or globally, consumers are the main beneficiaries of agricultural research because low price elasticity of demand for agricultural products means that higher productivity achieved by innovation will translate into lower prices (Guttman 1978). However, in a world of highly efficient global transportation, the relatively small share of benefits from lower prices that accrues to consumers in a single state tends to limit within-state consumer support for agricultural research that increases national or global productivity (Rose-Ackerman and Evenson 1985).

For region-specific production-oriented research (especially on crops as opposed to animals), the negative price response on international markets may be negligible. Local farmers tend to get a substantial share of the benefits of this type of research, given the level of other research activity. Political support tends to be high for region-specific innovation, suggesting that farmers have substantial influence on relevant research spending. Empirical studies have found that U.S. state spending on agricultural research significantly and positively correlates with state characteristics such as per capita income, the share of rural population, the number of large farms, the political influence of farmers, and the number of firms producing agricultural inputs, while spending is negatively influenced by the ability to adopt technology produced in other states (Rose-Ackerman and Evenson 1985; Guttman, 1978; Huffman and Miranowski 1981). In developing countries, on the other hand, underinvestment in regionally specific crops such as cassava and sweet potatoes in the 1970s relates to the relatively low influence of staple food producers and small farmers on research agendas (Judd, Boyce, and Evenson 1987).

2.4   Intellectual Property for Agricultural Innovations

2.4.1   Methods for Protecting Intellectual Property Rights in Agriculture

Beginning with the 1930 Plant Patent Act, the United States has created a number of institutional innovations in the form of intellectual property rights to accommodate the needs of different agricultural technologies. The eligibility criteria, duration, and nature of rights conferred vary across these IPRs, and a single innovation may be protected under multiple mechanisms under the same or different legal jurisdictions. The first IPR specifically for plants was introduced by the 1930 Plant Patent Act, allowing plant patents for new and distinct varieties of most asexually propagated plants, while the U.S. Department of Agriculture (USDA) administers the separate Plant Variety Protection Certificates (PVPCs) for sexually reproduced plant varieties, in accordance with the 1970 Plant Variety Protection Act (PVPA). United States legislators based the PVPA on the International Convention for the Protection of New Varieties of Plants (UPOV)—an agreement established in 1961 by a group of Western European countries that lays out a model system for plant breeders' rights, which was itself influenced by the United States Plant Patent Act of 1930.

Subsequently, the controversial ruling of the 1980 *Diamond v. Chakrabarty* case (*Diamond v. Chakrabarty* 447 US 303, 1980) confirmed the applicability of utility patents for living organisms in the United States (*Ex parte Hibberd* 227 USPQ 433 [Bd. Pat. App. & Int. 1985]). Since the decision, utility patents have been applied to plant varieties, genetically engineered organisms, processes for expressing transformations, genes, traits, and materials. The 2001 *J.E.M. v. Pioneer* Supreme Court case confirmed the legality of joint protection under utility patents and a PVPC or a plant patent (*J.E.M. AG Supply v. Pioneer Hi-Bred International* 122 S. Ct. 593, 2001).

Trade secrets, trademarks, and geographical indications (GIs) may also apply to agricultural innovations. Legally a right based on state common law, trade secrecy can be invoked by firms that can demonstrate sufficient efforts to protect proprietary information of commercial value. It is frequently relied upon by innovators prior to obtaining a patent or other IP protection (Friedman, Landes, and Posner 1991) but has high value independent of patenting; important surveys have shown that innovators in most industries rank secrecy higher than patenting for effectively appropriating commercial value (Levin et al. 1987; Cohen, Nelson, and Walsh 2000). The model trade secret law, known as the Uniform Trade Secrets Act, has been adopted by forty-six states. Trade secrecy has been more important recently in agricultural biotechnology research because advances in biotechnology have improved detection of infringement (Boettiger et al. 2004).

To the extent that bioenergy development relies upon plant innovation, we can expect most of the preceding mechanisms (and their drawbacks, discussed in the following) to be relevant for energy technology. A common area of confusion concerns the jurisdiction of IPRs. It is important to note that a patent or PVPC can be enforced only in the jurisdiction in which it is granted (Taylor and Cayford).[6] While the Trade-Related Aspects of Intellectual Property Rights agreement (TRIPs) among the World Trade Organization (WTO) members mandates minimal standards for all types of IPRs, only copyright has virtually global reach, under the Berne Convention for the Protection of Literary and Artistic Works, which is largely incorporated in the WTO TRIPs agreement.

2.4.2   The Private-Sector Response

The private sector response to the new opportunities and incentives in agricultural biotechnology has been focused and forceful. Since 1987, over 55 percent of all field trials for genetically modified (GM) crops have been on corn and soybean varieties. Global GM crop value and planted area is almost entirely in soybeans, corn, cotton, and canola (James 2008; Runge and Ryan 2004). Monsanto has become the dominant firm in generation and diffusion of agricultural biotechnology, concentrating its activities on corn, soybeans, and cotton incorporating herbicide resistant "Roundup Ready" technology and pest-resistant traits based on crystal proteins derived from samples of Bacillus thuringiensis (Bt) (Lemarié 2001). It is unlikely that public research institutions could have matched the efficiency and scale of Monsanto and other leading firms in these activities. For example, in 1998,

---

6. The well-publicized case of "Golden Rice" transformed to include provitamin A, has been characterized as subject to scores of widely held patents and as an excellent example of private-sector collaboration in licensing these patents for use in poor rice-consuming countries. In fact, few if any of the patents were relevant to production or use of the technology in major poor rice-consuming countries; licenses related mainly to material transfer agreements. (See Binenbaum et al. 2003).

Monsanto spent $1.26 billion in agricultural biotechnology R&D, eclipsing the total CGIAR investment of $25 million in agricultural biotechnology that same year (Pardey and Beintema 2001).

2.4.3 IPR Limitations and Drawbacks

While the most popular modern rationale for granting IPRs is that they motivate technological innovation and dissemination by at least partially privatizing associated social benefits, the experience in agriculture shows both the strengths and the limitations of IPRs in a well-balanced system of dynamic, creative research. First, private research has not been (and likely never will be) a complete substitute for public agricultural research (Alston, Norton, and Pardey 1995). While the recent strengthening of intellectual property rights in the United States opened the way for increased private participation in basic plant breeding research by major firms such as Monsanto and Pioneer (Falck-Zepeda and Traxler 2000), the private sector primarily performs applied research with the goal of producing a profitable technology or product. Private-sector research critically depends on the public sector to produce the "building blocks" for technology (Alston, Pardey, and Roseboom 1998; Stokes 1997; Huffman and Evenson 2006; Just and Huffman 1992).

Economists have recognized that the intellectual property protection of research at public institutions might erode the provision of such public goods. For example, Just and Huffman (2009) note that while the 1980 Bayh-Dole Act led to a jump in university patenting (from less than 400 patents per year before Bayh-Dole to 1,100 per year by 1989 and over 3,000 per year in the 1990s), the act may have reduced the pool of basic research supporting private research, shifting the focus of public research toward shorter-term incentives such as patents, at the expense of future public goods research.

Furthermore, there appears to be great variation in private responses to different forms of intellectual property. While Fuglie et al. (1996) found that in the United States, private investments in agricultural biotechnology increased fourfold (nominal) in the first twelve years after the *Diamond v. Chakrabarty* ruling, Alston and Venner (2002) were unable to find evidence of increased private-sector investments in wheat breeding as a response to the PVPA, which they conclude is more relevant to marketing than research.

2.4.4 Anticommons in Agricultural Research and Market Concentration

There is mounting evidence that multiple, mutually blocking intellectual property claims on inputs are hindering access to research tools that can be incorporated in the marketed products of agricultural research (Wright and Pardey 2006; Pardey et al. 2007). The rising application of IPRs to plant components and processes imposes high transaction costs for researchers who must acquire or license fragmented proprietary inputs to develop and commercialize a single downstream innovation.

Agricultural economists have long been concerned that patents on locked-in but otherwise noncrucial genetic technologies have been retarding innovation and affecting the market structure of private research. In the field of plant biotechnology in particular, where ownership of the genes, markers, or promoters incorporated in a single innovation is fragmented, upstream IPR-holders, unwilling to allow commercialization of varieties using their property, have in some instances foreclosed university development of new crops or technologies.[7] The broader economics profession has become focused on these issues more recently, due to growing problems with blocking patents on embedded software (Bessen and Hunt 2003; Lemley and Shapiro 2005, 2007; Cohen, Nelson, and Walsh 2000). The Supreme Court appears to have acknowledged this problem in its eBay decision (*eBay Inc. v. MercExchange, L.L.C.,* 547 U.S. 388 [2006]), which reduced the ability of holders of blocking patents to use the threat of injunction to extract high royalties from infringers.

In a 2003 *Science* article signed by fourteen university presidents, chancellors, and foundation presidents, the authors highlight the negative effect of intellectual property rights on "freedom to operate" in agricultural research (Atkinson et al. 2003). In fact, universities themselves appear to have contributed to the problem by insisting on the use of material transfer agreements (MTAs) governing exchanges of research materials between researchers, to protect university intellectual property rights and limit university liability.

A recent survey indicates that, for scientists focused on their own research goals rather than commercialization of innovations, the problems associated with MTAs are more salient than concerns about patent infringement. Over a third of agricultural biologists at land grant universities reported delays in obtaining access to research tools in the five years preceding the survey, with a mean of two delays and a mean duration of over eight months (Lei, Juneja, and Wright 2009). They attributed the vast majority of these delays to problems experienced by university administrators in negotiating MTAs. Researchers reported responding to hold ups by substituting tools, sometimes of lower efficacy, and, in some cases, by abandoning the project altogether. Although a substantial portion of the sample were patentees, most respondents view intellectual property protection as a net negative for progress of research in their fields.

7. Examples are the cases of the genetically engineered (GE) tomato and herbicide-resistant strawberry at the University of California at Berkeley, an herbicide tolerant barley, and an herbicide tolerant turf grass at the University of Michigan (Wright 1998; Wright et al. 2007; Erbisch 2000). More recently, commercialization of transgenic hypoallergenic wheat technology has been blocked by a combination of patent protection and regulatory costs (P. Lemaux, personal communication). Generation of further examples is unlikely, given that independent university implementation of new transgenic technologies is widely regarded as economically infeasible due to the combined effects of blocking patents and regulatory costs and delays.

   Follow-up interviews revealed that scientists view university administrators as principally concerned with protecting their institutions' financial interests, including protecting claims to potential intellectual property value and reducing exposure to potential liability. Indeed, Glenna et al.'s (2007) survey reveals that land grant university administrators, on average, rate the provision of new funds and research support as the greatest advantage of university-industry research collaborations but believe that scientists choose projects based on research enjoyment and scientific curiosity.

   Private firms in agricultural biotechnology, unlike bench scientists, cannot ignore infringement issues. Often, firms have to commit to commercializing a product involving many patents, subject to time-consuming development and regulatory approval processes, and thus exposed to hold up even by owners of nonessential technologies that become "locked in" as innovation progresses. Firms have established freedom to operate largely by merger and acquisition of patentees rather than by in-licensing.

   Since the mid-1990s, a relatively dramatic series of industry purchases led to concentration of the agricultural biotechnology sector. As a result, a relatively modest number of private hands controls a major portion of patents in the field (Murray and Stern 2007; Marco and Rausser 2008). In effect, the threat of an anticommons with many parties capable of stopping a line of business in progress via an injunction has been avoided by a strategy that does not rely solely on in-licensing, but instead deals with potential hold ups by acquisition or merger. The European conglomerate AgrEvo acquired Plant Genetic Systems (PGS) for $730 million, $700 million of which accounted for the estimated value of PGS-owned patents on plant traits. Perhaps more extreme was Monsanto's 1997 purchase of Holden's Foundation Seeds for $1.1 billion, when Holden's gross revenues were just $40 million. In 1998, Monsanto paid $2.3 billion to acquire the 60 percent of DeKalb Genetics Corporation, which it did not already own. DeKalb, the owner of a major corn transformation technology, had a 1997 total revenue of $450 million (United States Department of Justice 1998; Marco and Rausser 2008). The following year, DuPont purchased the remaining 80 percent of Pioneer Hi-Bred International that it did not already own for $7.7 billion (Marco and Rausser 2008).

   Based on USDA data between 1988 and 2000, Brennan et al. (2005) found that when including mergers and acquisitions, the top ten firms (rated by number of patents held) owned more than half of the agricultural biotech patents granted through 2000, whereas if patents acquired via purchases and mergers are excluded, the share owned by the top ten firms would be only one-third. One firm (Monsanto) currently accounts for almost two-thirds of all public and private plant biotech field trials in the United States. King and Schimmelpfennig (2005) show that by 2002, six firms (Monsanto, Dow, DuPont, BASF, Bayer, and Syngenta) controlled over 40 percent of the agricultural biotech patents owned by the private sector, with the subsidiaries

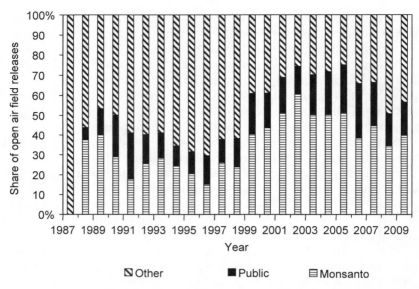

Fig. 2.1   Yearly shares of genetically engineered organism controlled open-air field releases

Source: Source data from U.S. Department of Agriculture, Animal and Plant Health Inspection Service. http://www.aphis.usda.gov/biotechnology/status.shtml.

acquired by these firms through mergers and acquisitions responsible for 70 percent of their total patent stocks. Concentration in the innovation market has continued to increase; USDA data on field releases of new genetically modified organisms show Monsanto's dominance in the testing of new GM products (figure 2.1).

Thus, the need to ensure freedom to operate in an environment of initially fragmented and decentralized proprietary claims enhanced the normal tendency of firms in a new industry to consolidate in order to avoid transaction costs, exploit increasing returns to scale, and establish market power (Rausser 1999). Notably, Marco and Rausser (2008) empirically show that in the plant biotechnology industry, the enforceability of a firm's patent portfolio is a good predictor of participation in consolidation. Additionally, the authors note that a number of mergers, including Monsanto-Calgene and Monsanto-DeKalb, occurred in the context of patent infringement litigation.

It appears to be true that IP-induced consolidation in agricultural biotechnology is negatively affecting the very same innovation and dissemination incentives that justified IPRs in the first place. By blocking new firms from entering, market consolidation may suppress future innovation (Barton 1998; Graff et al. 2004). Lack of freedom to operate particularly affects nonprofit research institutions, which cannot solve the problem by merger or acquisition of blocking firms.

2.4.5   Attempts to Alleviate the Anticommons: BiOS and PIPRA

Early recognizing the threat from lack of freedom to operate, public and nonprofit parties have formed institutions that endeavored to construct alternative technologies that could circumvent patents blocking key transformation technologies. These include the Biological Innovation for Open Society (BiOS) and Public-Sector Intellectual Property Resource for Agriculture (PIPRA) initiatives, which have both attracted widespread attention from biologists in the biomedical sector.

Inspired by the successful open source models in software, BiOS, an open source initiative arising out of the nonprofit biotechnology firm CAMBIA, was formed to license rights to use protected innovations, including those that could substitute for technologies protected by blocking patents, in exchange for a commitment to share any downstream technologies with all BiOS members (Jefferson 2006). It appears that this bold initiative has not functioned as anticipated. Participants willing to access CAMBIA technologies offered by BiOS are apparently reluctant to follow the lead of CAMBIA and offer their own technologies under an open source license.[8] A more fundamental problem might be that any success achieved in open source software using copyright licenses to prevent appropriation of the core technology might be difficult to replicate in a system that relies on patent protection (Boettiger and Wright 2006).

Founded with support from the University of California and the Rockefeller and McKnight Foundations, PIPRA was intended to act as a clearinghouse for information about patenting and licensing of technologies originating in the public and nonprofit sector. The goal was to facilitate commercialization and adoption of new technologies in less-developed countries and of "minor" crops such as the fruits and vegetables produced in California. The common problem of both target groups was that their markets were too small to attract much interest from the major agricultural biotechnology corporations. The Public-Sector Intellectual Property Resource for Agriculture's intellectual property strategy, similar to that of CAMBIA, was to protect proprietary claims for commercial use in developed countries (consistent with federal policy expressed in the Bayh-Dole Act of 1980 and with the aims of university licensing objectives), while simultaneously providing access to users in less-developed countries and producers of minor crops and reserving public and nonprofit institutions' rights to use the inventions in developing-country applications.

While these initiatives yielded notable scientific advances (see for example Broothaerts et al. 2005), neither has yet furnished an efficient and completely unencumbered biotechnology package that is a good substitute for proprietary blocking technologies that received wide market acceptance. Indeed,

8. Stricly speaking, the license was not open source as participants are charged on a sliding scale based on size.

the experience of these initiatives constitutes strong evidence of the blocking capacity of proprietary claims over key elements of plant transformation. The lack of attractive unencumbered alternative technology sets may be the main reason why efforts to encourage collaboration in open source type ventures have not made more headway. However, had they already succeeded at this level, the regulatory hurdles would still have loomed large.

Although attempts to offer unencumbered alternatives to key technologies are continuing, both CAMBIA and PIPRA are currently emphasizing provision of easily accessible information to researchers in developing countries and nonprofit institutions regarding freedom to operate within the context of patenting as the dominant paradigm. They offer other services that can assist public and nonprofit research institutions in navigating patent thickets and identifying intellectual property issues before they become serious problems. The Public-Sector Intellectual Property Resource for Agriculture provides educational and informational services to facilitate navigation of the IP landscape and promote innovation-enabling collaboration as well as a valuable IP handbook. CAMBIA offers, among other services, its "Patent Lens" to provide accessible guidance as to the patent landscape relevant to the plans of researchers in biotechnology (Graff et al. 2003; Atkinson et al. 2003; Delmer et al. 2003).

2.5   Public-Private Collaborations

Channeling fruitful basic research from the public sector into applied research efforts by the private sector is a key, though problematic, step in the innovation "pipeline." Economists have often highlighted the potential for public-private collaborations to bridge this gap and have argued for their critical role in spreading agricultural biotech innovations for consumers in developing countries (Rausser, Simon, and Ameden 2000; Byerlee and Fischer 2002; Tollens, Demont, and Swennen 2004; Parker and Zilberman 1993). In light of slowed federal research support between 1980 and 2000, collaborations with the private sector have become increasingly attractive for public-sector researchers. However, public-private collaborations do not currently account for a large portion of agricultural research in practice. A few examples illustrate of the potential benefits and drawbacks of these agreements.

Overcoming conflicts between private-sector interests and public responsibilities is a key challenge to fruitful public-private collaboration. In their 1986 analysis of the Canadian malting barley industry, Ulrich, Furtan, and Schmitz (1986) find that the availability of grants from and collaborative work with the brewing industry led the public sector to pursue traits to improve malting quality. Although the research yielded both public and private benefits, the collaboration siphoned public resources away from yield-related traits. Ulrich and colleagues calculate social gains would have been 40 percent higher or more had public researchers focused only on yield-related traits. Note that this example is drawn from an era when intellectual property protection had yet to strongly influence the direction of research activity in agriculture.

The subsequent proliferation of patenting in universities has increased the potential for diversion of the university agenda toward private monetary returns and away from the direct public interest. Such a real or apparent shift in focus of public institutions away from public goods research might further reduce public support for universities (Just and Huffman 2009), even if the ultimate motive is to ensure survival of the institution in an era of public cutbacks. Supporting this argument are Glenna et al.'s (2007) findings that land grant university administrators rate further support for research and increased funding as the main benefits from university-industry collaborations and potential for conflicts of interest as their main drawback.

An innovation in private support for public agricultural research was established by the 1998 agreement between the University of California at Berkeley and the Novartis Foundation. The latter was selected by Berkeley as partner after competitive bids from several firms. The agreement generated much opposition, motivated by opposition to the genetic transformation technologies involved in the research to be funded, by objections to the decision-making procedures that led to the partnership, and by concern that the university was selling its plant biology agenda to the highest bidder and that new private-sector influence might distort the direction of university research. The Novartis Foundation had the right to a vaguely specified subset of the research results in plant and molecular biology and had a minority of seats on the committee allocating project resources (Busch et al. 2004; Rausser 1999).

In hindsight, the $25 million received from the Novartis (later Syngenta) Foundation appears to have been associated with an increase in the level and the diversity of funding of Plant and Molecular Biology at Berekeley, with little if any affect on research direction. It appears that no valuable patents were obtained by the funder. Indeed, a similarly generous arrangement is unlikely to be achieved in future public-private partnerships in biotechnology. Nevertheless, an ex post review (Busch et al. 2004) concluded, among other things, that the agreement did affect the processes leading to the initial denial of tenure (later reversed) to a prominent critic of the agreement. Even relatively hands-off funders, it seems, create the hazard of a threat to academic freedom in the university.

Subsequently, Berkeley has become the lead partner (with Lawrence Berkeley National Laboratory and the University of Illinois at Urbana-Champaign) in the $500 million Environmental Biosciences Initiative (EBI), funded by British Petroleum and aimed at developing means of converting cellulosic biomass to a substitute for petroleum. Some of the intellectual property provisions of this agreement are more favorable to the funder than in the Berkeley-Novartis agreement, but the EBI has not generated the same degree of opposition on campus.

In another model of public-private collaboration known as Cooperative Research and Development Agreements (CRADAs), the United States provides research funds to national laboratories contributing nonfinancial resources in order to produce a commercial technology in collaboration with a private firm. Any proprietary material may be owned by both parties, but the private collaborator gets priority in licensing (Day-Rubenstein and Fuglie 2000). Since 1987, the USDA has formed at least 700 CRADAs with private firms. The CRADAs have produced and commercialized at least a handful of important innovations (Day-Rubenstein and Fuglie 2000).[9] In accordance with the goal of connecting the basic and applied ends, CRADA research focuses on a "middle ground" between public and private goods (Day-Rubenstein and Fuglie 2000). However, concerns that CRADAs and other public-private collaborations divert funds from public goods or basic research to more applied research highlight the need to carefully design collaboration agreements and to consider the unintended effects of undertaking collaborations (Just and Rausser 1993). In addition, claims that CRADAs provide unfair advantages for the private partner (even if, as in the case of Taxol, the contract was awarded by auction) illustrate that any agreement mechanism providing profits high enough to incentivize private investment in a risky enterprise will be met with critical political comment if the CRADA succeeds.[10]

Public-private collaborations also tend to be a target of public scrutiny that can negatively influence the credibility of government or other public entities to uphold their public responsibilities. In particular, public-private collaborations that do not convincingly protect public decision making from improper influence weaken the regulatory role of government. For example, widespread public distrust of government's ability to regulate genetically modified organisms in Europe must be considered in the context of several major government failures to protect consumer safety. In particular, concerns that governments might prioritize industry welfare over public safety were fueled by poor regulation both paving the way for and in response to incidents such as the Chernobyl disaster and the spread of bovine spongiform encephalopathy (BSE), commonly known as "mad cow disease" (Park 1989; BBC News 2000; CNN 2001).

9. Most notably, a CRADA is responsible for production of the anticancer drug Taxol, based on the bark of the Pacific yew tree, and involving the USDA and its Forest Service (Koo and Wright 1999). This highly successful drug is used in treatment of ovarian and breast cancers. In the agricultural field, CRADAs have been associated with the development of a number of pest and disease controls, a chicken vaccine for Marek's disease, and a chemical compound to reduce soil erosion (Day-Rubenstein and Fuglie 2000).

10. Even prominent economists will make such comments. See, for example, the claim by Boldrin and Levine (2008, chapter 1) that any returns above "break-even" were superfluous to the incentive needed for Boulton and Watt to produce their celebrated steam engine.

On the other hand, research consortium models such as those adopted by the Latin American Maize Project (LAMP) and the Germplasm Enhancement of Maize Project (GEM) have been lauded for productively balancing public goods research with commercial viability (Knudson 2000). Initiated by then Pioneer CEO William Brown, LAMP was established in 1987 as a cooperative effort to regenerate and utilize maize landraces, supported by funding from Pioneer and resources from the United States, eleven Latin American countries, and CIMMYT (Knudson 2000; CIMMYT 1997). Of the 12,000 accessions evaluated by LAMP, 51 accessions plus 7 donated from DeKalb formed the source material for the USDA-ARS GEM Project. Jointly funded by an array of private, public, and nonprofit collaborators and utilizing research from federal programs, state programs, and private industry, GEM's structure recognized needs from the local to national levels and maintained industry support through appropriate IPR provisions (Knudson 2000).

2.6   Government Regulation

From an economic perspective, government regulation is necessary to correct for externalities such as those related to environmental quality, food safety, and public security. Aside from immediate social, environmental, and economic costs, past regulatory disasters and their aftermaths demonstrate that ineffective or disingenuous government practices create a secondary problem of reduced public confidence and support. On the other hand, the costs of regulatory compliance can discourage innovation and technology adoption. The challenge to government is to strike a balance between supporting technological improvement and protecting public safety.

2.6.1   Public Trust

Experience in agriculture demonstrates that a poor track record or a handful of extreme cases may durably erode public confidence in the government's ability to prudently select and monitor new technologies. The U.S. Three Mile Island incident and the Ukrainian Chernobyl disaster caused significant curtailment of nuclear power projects in Sweden and Italy, respectively, and weakened public trust in government officials and scientists alike (Weingart 2002). Studies of government regulation and public perceptions drew parallels between the cases of nuclear power and agricultural biotechnology (Sjöberg 2001; Poortinga and Pidgeon 2005). Indeed, Chernobyl's contamination of European foods was initially denied by many governments, and Chernobyl's effect on public perceptions of technological risks, government competence, and the reliability of public assurances has contributed to the prohibition of GM foods in Europe (Vogel 2001; Lusk, Roosen, and Fox 2003).

Past crises in food safety have also demonstrated the consequences for public health, government reputation, and industry. After the BSE outbreak was recognized in Britain, European regulatory systems were further discredited by the British Ministry of Agriculture's insistence that BSE posed no threat to human health and the French government's slow response to BSE (Vogel 2001; Daley 2001). The direct effects of BSE on beef consumption and the subsequent trade bans caused heavy losses to beef producers. The containment costs borne by the British government reached £700 million per year in 1996, or about 0.1 percent of gross domestic product (GDP) (Gollier, Moldovanu, and Ellingsen 2001). Another example is the release by the European multinational firm Aventis of genetically engineered corn, StarLink, into the U.S. food supplies, when the corn was approved only for feed. Fortunately, no health damage to consumers was detected. Aventis agreed to pay $110 million plus interest to farmers whose crops were contaminated (O'Hanlon 2004; Pollack 2001; Shuren 2008) to compensate for adverse effects on prices of exports due to fears of harmful contamination.

The StarLink case not withstanding, the lack of major food-related regulatory disasters in recent U.S. experience likely has contributed to greater public acceptance and employment of agricultural biotechnology in the United States relative to Europe (Vogel 2001; Nelson 2001). Note that the effects on public perception are not confined by national boundaries and can become confounded with strategic market maneuvers. For example, China changed its plans to approve GM foods when the StarLink incident cooled demand for GM corn by importers Japan and South Korea (Cohen and Paarlberg 2004). In North America, the U.S. wheat industry compelled the agricultural biotech firm Monsanto to release the first genetically engineered spring wheat in the United States and Canada, or not at all. The technology was shelved in both countries when Canada rejected the product (Berwald, Carter, and Gruère 2006).

Considering the commonalities of the preceding cases, we would expect public trust problems to be most influential for technologies related to foods, with uncertain but potentially widespread and irreversible risks, and in cases where the benefits are dubious from the consumer viewpoint (Arrow and Fischer 1974; Brush, Taylor, and Bellon 1992). Bioenergy products, especially if they are not related to foods or feeds, might be better accepted due to the widespread recognition of the benefits of a cleaner and more secure energy supply.

2.6.2   Nonadaptive Regulatory Systems and Unintended Consequences

While negligent regulatory systems can precipitate public safety disasters, the costs of extremely restrictive regulatory systems are less transparent but may also be severe. They can impede the application of technology, implying missed opportunities to realize the benefits of research and innovation. Cases in various countries show that both lack of regulatory capacity and public distrust may lead to regulatory procedures that slow adoption of technology. In addition, relatively stringent regulatory standards for a particular set of technologies create an advantage for substitutes.[11] Thus, government safety standards have indirect but important effects on market structure.

The case of U.S. agricultural chemical regulation illustrates some of the trade-offs related to standard setting. Ollinger and Fernandez-Cornejo (1995) found that between 1972 and 1989, in response to three amendments tightening regulatory requirements, industrywide research spending increased, but the share of regulatory costs in total R&D increased from 18 percent to an astounding 60 percent. The authors also found that each 10 percent increase in regulatory costs corresponded to a 2.7 percent reduction in the number of new pesticides products but also decreased the negative environmental qualities of the pesticides produced. Regulatory costs affected the market by increasing foreign capture of market share, reducing the number of small firms and broadening opportunities for biological pesticides and genetically modified organisms (GMOs). As this example shows, the net effects of regulation are hard to pin down.

Many governments have recently enacted regulatory standards for the release of agricultural biotechnology to address concerns about potential ecological and food safety risks. While some scientists fear the standards for transgenic crops do not adequately inform us about potential risks, economists have argued that these regulations hinder implementation of important technologies (Fuglie et al. 2006; Zilberman 2006; Pardey et al. 2007; Cohen and Paarlberg 2004). For example, lack of mutual recognition of regulatory standards and test results requires duplication of field trials for transgenic crops in some East and South African countries, without generation of new information between test trials, and has slowed commercialization (Pardey et al. 2007; Thomson 2004).

In India, regulatory authorities approved the first field trials for Bt transgenic cotton in 1998. Although the crop had been grown as early as 1996 in countries such as the United States, Australia, South Africa, Mexico, Argentina, and China, outcry by nongovernmental organization (NGO) groups claiming to represent farmers influenced regulatory officials to delay final approval for Bt cotton until 2002—about six months after authorities discovered some 500 Gujarati farmers had already illegally planted Bt cotton seeds. Public plans to destroy the standing crop were abandoned after demonstrations by thousands of farmers. Even more extreme is the case of a transgenic mustard variety that underwent field trials in India for at least nine years (Cohen and Paarlberg 2004) but is yet to be commercialized. If biofuel technology utilizes transgenic crop innovations, inflexible and

11. Graff and Zilberman (2007) argue that the interests of European pesticide and herbicide producers have influenced the development of negative attitudes on that continent to agricultural biotechnology.

inefficient regulatory systems could significantly defer this alternative fuel's production and utilization.

For transgenic food crops in the United States, the Food and Drug Administration, Environmental Protection Agency, and USDA Animal and Plant Health Inspection Service all have regulatory authority concerning potential risks related to food safety, pesticidal properties (if applicable), and field testing, respectively (Vogt and Parish 1999; Fuglie et al. 1996). While public opinion and government policy create a less hostile environment for GMOs relative to the regulatory environments in areas such as Europe and Africa, some economists maintain that U.S. regulation for transgenic crops is unduly restrictive, relative to regulations for competing technologies (Miller and Conko 2000).

In industries subject to market concentration, such as the agricultural biotech industry, insights from capture theory suggest pushes for regulatory reform are likely not forthcoming. High compliance costs associated with stringent regulatory standards may even be preferred by firms with market power because such standards create barriers to entry and build consumer approval (Zilberman 2006). For example, in 1925, California adopted the One-Variety Cotton Law, requiring nearly all California cotton growers to plant a single USDA-controlled cotton variety in the interests of quality control. In their 1994 analysis of the regulation, Constantine, Alston, and Smith (1994) argue that any initial benefits were lost over time due to unforeseen technological changes, institutional reforms, and regulatory cost increases. While the industry was partially deregulated in 1978, private beneficiaries of the regulation successfully resisted further reform for a number of years, despite the regulation's large social costs.

Economists have critiqued the precautionary principle (cited by European GMO regulators) for failing to respond to new information (Gollier, Moldovanu, and Ellingsen 2001; Vogel 2001). In agricultural biotechnology, the firm bears the costs of regulatory compliance. Although farming of biotech food crops has grown rapidly in recent years, the still narrow breadth of the market may be influenced by regulations slowing commercialization of new transgenic traits beyond herbicide tolerance and Bt-based insect resistance in corn, soybeans, canola, and cotton (Pardey et al. 2007).

2.7   Technology Adoption

Timely adoption is necessary in order to realize the benefits of innovation. The most relevant aspect of adoption in the agricultural sector is the dispersed application and management of technologies adapted to local environments. The period between technology development and widespread adoption by farmers can be quite long (and infinite for technologies that never take hold). Griliches's (1957) work on hybrid corn adoption and subsequent empirical studies showed that the rate of technology diffusion increased with profitability from its adoption.[12] A multitude of economic studies has taught us that factors such as risk and heterogeneity are crucial determinants of adoption. In addition, the canon of relevant political economic work highlights the dependence of adoption on consumer preferences, political interests, and the appropriateness of technology. With respect to energy innovations, we might expect analogous issues to arise for technologies with similarly dispersed applications. Again, production of biofuel crop technologies is a clear example, but prior adoption research might also be relevant for technologies like cookstoves for less-developed countries, mini or micro hydroelectric systems, wind farms and building efficiency innovations, all of which could well have user-dependent outcomes affected by the heterogeneity of regulations and of the natural environment.

2.7.1 Profitability and Heterogeneity

The recent increases in market concentration due to strengthened IPRs in agriculture (as discussed in the preceding) might well imply oligopoly pricing that reduces adoption of technologies integrating proprietary components. However, the dispersed use of agricultural technologies in heterogeneous environments constrains the extent of noncompetitive pricing if spatial price discrimination is costly. Thus, even firms exercising market power may have to employ low-price policies in order to encourage rates of technological exposure and adoption that make the discounted value of the innovation positive and attractive to investors. The empirical study by Falck-Zepeda et al. (2000) on markets for transgenic soybeans and cotton estimated that firms Monsanto and Delta and Pine Land (D&PL) adjust price to keep about 21 percent of the benefits generated from these innovations, while about 59 percent of these benefits flow to farmers, despite essential monopoly power by Monsanto and D&PL. A follow-up study by Oehmke and Wolf (2004) estimated heterogeneity in technology adopters to account for 80 percent of farmer rents. Although the question has yet to be addressed seriously, it appears that if "degraded" lands are favored for biofuels production due to low carbon release upon cultivation, their heterogeneity could well reduce the speed of technology adoption, thus reducing monopoly rents and perhaps investment in optimizing the technology.

When the newest technology is a substitute for their current technology, farmers will not adopt the new technology unless the net benefits of switching are positive, leading to an additional constraint on oligopoly pricing for the new technology (Pray and Fuglie 2001). If the current rate of innovation is rapid, the loss of the option to wait and use an even better prospective technology may increase the cost of adopting the current best technology when the sunk costs of learning or complementary investment are significant.

A large body of research on heterogeneity and adoption demonstrates

12. Feder, Just, and Zilberman (1985) provide a summary of empirical work.

that a number of other factors moderate the profitability of adoption at a given time. While a full review is beyond the scope of this chapter, we briefly provide a few examples to illustrate the diversity of this work. Studies such as those by David (1969), Feder (1980), and Ruttan (1977) discuss the influence of farm size (perhaps as a proxy for wealth or increased access to credit, information, or production inputs) on the rate of technology diffusion or level of individual adoption for various types of technology and institutional frameworks (Feder, Just, and Zilberman 1985; Qaim et al.). The findings from a recent paper by Fernandez-Cornejo, Hendricks, and Mishra (2005) imply that small U.S. farms that supplement farm income with off-farm activities are more likely to adopt time-saving technologies such as herbicide-tolerant crops or conservation tillage. Examples of research evaluating the effect of heterogeneity in land quality on adoption decisions include Caswell and Zilberman's (1986) analysis of water-holding capacity and irrigation technology and Rahm and Huffman's (1984) article relating soil quality to adoption of corn varieties. Recently, access to information and social capital have been highlighted as determinants of adoption of crop varieties such as hybrid corn and wheat, mechanical innovations such as tractors, and livestock technologies, to name a few (Skinner and Staiger 2005; Matuschke, Mishra, and Qaim 2007; Abdulai and Huffman 2005). Finally, a plethora of research discusses the role of risk and uncertainty associated with adopting new technologies.

Another important insight is that technologies that reduce input use per unit output may, via price effects, increase total input demand with respect to the efficiency-adjusted unit, as observed in studies of water or energy use in modern irrigation technology (Caswell and Zilberman 1986). Thus, yield-enhancing varieties of biofuels crops may increase, rather than reduce, the area planted to such crops.

2.7.2   Policy Implications

Given the heterogeneity of the myriad of moderating factors, the adoption of agricultural technologies over extensive geographic areas is enhanced by directed local efforts and adaptations (Knowler and Bradshaw 2007). Key decisions for government are whether and how to alleviate risks and high fixed costs associated with adoption. The relevance of risk is dependent upon farmer perceptions and the type of technology (for example, see Fernandez-Cornejo, Beach, and Huang 1994). The adoption of crop technologies is scale-neural in the sense that farmers can test the new crop on a small portion of land, thus reducing risk problems and allowing farmers to learn through use (Feder 1982; Feder, Just, and Zilberman 1985). This strategy might be less relevant for biofuels if their introduction depends upon a large investment in local processing facilities. In this case, extra attention should be paid to spatially dispersed adaptive and evaluative research (Judd, Boyce, and Evenson 1986) and appropriate extension services, public or private.

Agricultural extension systems are designed to provide farmers information about new technologies and thus facilitate technology transfer or adoption. Researchers have estimated high rates of return to extension work in the United States and have demonstrated that, provided there are attractive technologies awaiting adoption, contact between farmers and extension agents promotes technology adoption in some less-developed countries (Huffman and Evenson 2006; Abdulai and Huffman 2005; Polson and Spencer 1991).

In industries requiring high up-front investments in infrastructure or regulatory compliance, adoption may be retarded by lock-in of old technologies. In agricultural biotechnology, regulatory requirements for testing a new variety, and fragmented IP claims on inputs for a single new technology, create high costs of entry for followers once a leading innovation has become approved established. Private firms owning IP might raise their prices for access to proprietary technology after strategically pricing low to induce diffusion and dependence.[13] This issue is akin to a problem of patented technology incorporated in regulatory standards, familiar from the literature in electronics and communications technology, although the connections between the relative literatures have not, as far as we are aware, been fully explored.

Note, however, that if farmers perceive a technology to be extremely attractive or to provide significant benefits, they will not only adopt at impressive rates but may also perform adaptive innovation themselves, regardless of the policy regime in place. A case in point is the exchange and development of unapproved genetically modified seeds by Gujarati farmers and the success of these farmers' opposition to federal attempts to destroy the unapproved standing crop (Pray et al. 2005). The case of "no-tillage" agriculture in the United States, one of the prominent agricultural innovations in the late twentieth century, is also a striking example of users innovating and adopting in response to prices and practical environmental problems. The first no-till planting is credited to Kentucky grain grower Harry Young, an extension specialist who successfully applied his knowledge of scientific trials for new herbicides to develop this new method of crop growing in 1962 (Coughenour and Chamala 2001). Due to such benefits as increased yield and reduced requirements for labor, water, fuel, and machinery, the word about no-tillage spread rapidly between farmers, who traveled from neighboring states to learn about Young's technique. The ensuing years saw

13. For example, in the years before United States patents were generally published eighteen months after application, Monsanto encouraged widespread use of their 35S promoter in plant transformation. The fact that Monsanto had a patent on the technology was revealed only after 35S had been diffused widely, when the patent was granted and published. Innovators could commercialize their technologies incorporating the promoter only if they had a license from Monsanto because switching promoters would have required producing new transgenic technologies using a noninfringing promoter, followed by transformation of relevant cultivars, testing, and dissemination (Joly and de Looze 1996), an alternative so time-consuming as to preclude serious consideration.

rapid diffusion of the technique between farmers, who adapted the technology for other crops such as soybeans.[14] In view of the many complex legal and institutional innovation systems developed by governments, these examples are important reminders of the time-honored role of practitioners participating in technological development, with motivations ranging from necessity to curiosity.

2.8   Conclusion

In both the energy and the agricultural sectors, the demand for the services and goods produced extends across all the populated areas of the world, where heterogeneity in consumer preferences, environmental conditions, economic structures, and governmental policies mediate the applicability and appropriateness of innovations. For agriculture, the need for local adaptive technology has created global dispersion of research efforts and support. However, investments are by no means evenly dispersed across the globe. Instead, the majority of global spending comes from large countries that can internalize the benefits of research. The high rates of return to agricultural research in the United States were achieved through sustained public investment oriented by a clear mission. The establishment of institutional mechanisms in the nineteenth century created a public commitment to agriculture, but the extent and direction of public investments has been and will continue to be influenced by interest groups operating at multiple geographic scales.

The experience in agriculture should be of interest to those assessing the appropriate roles of public and private research. In agriculture, (a) IPRs have not strongly encouraged the private production of basic, essential research that is risky and often only pays off in the long run, (b) IPRs on key research inputs can impede freedom to operate in public research, and (c) IPRs on research inputs have led to market concentration and price markups, which should discourage or delay adoption. However, it appears that leading firms in the private sector have been particularly efficient in developing, promoting, and disseminating commercial technology packages, relative to what one might reasonably expect of a typically competent public or nonprofit entity, especially in the context of a disruptive and controversial technology.

For governments, experience in agriculture illustrates the importance of effective consumer protection, administration of standards and grades, and antitrust. With respect to innovations, governments face the challenge of balancing public trust and safety with exploiting the advantages produced by a changing technological environment. Critiques of the shortcomings of regulatory performance in the United States and elsewhere draw attention to

---

14. By 1970, 35 percent of farmers in southwest Kentucky had tried no-tillage for corn (Coughenour and Chamala 2001).

the need for further scientific study directed to the achievement of effective and dynamically adaptive regulation.

References

Abdulai, Awudu, and Wallace E. Huffman. 2005. The diffusion of new agricultural technologies: The case of crossbred-cow technology in Tanzania. *American Journal of Agricultural Economics* 87 (3): 645–59.

Adelaja, Adesoji O. 2003. The 21st century land grant economist. *Agricultural and Resource Economics Review* 32 (2): 159–70.

Alston, Julian M. 2002. Spillovers. *Australian Journal of Agriculture and Resource Economics* 46 (3): 315–46.

Alston, Julian M., Matthew A. Anderson, Jennifer S. James, and Philip G. Pardey. 2010. *Persistence pays: U.S. agricultural productivity growth and the benefits from public R&D spending.* New York: Springer.

Alston, Julian M., Connie Chan-Kang, Michele C. Marra, Philip G. Pardey, and T. J. Wyatt. 2000a. A meta-analysis of the rates of return to agricultural R&D: *Ex Pede Herculem?* IFPRI Research Report no. 113. Washington DC: International Food Policy Research Institute.

Alston, Julian M., Michele C. Marra, Philip G. Pardey, and T. J. Wyatt. 2000b. Research returns redux: A meta-analysis of the returns to agricultural R&D. *Australian Journal of Agricultural and Resource Economics* 44 (2): 185–215.

Alston, Julian M., George W. Norton, and Philip G. Pardey. 1995. *Science under scarcity: Principles and practice for agricultural research evaluation and priority setting.* Ithaca, NY: Cornell University Press.

Alston, Julian M., Philip G. Pardey, and Johannes Roseboom. 1998. Financing agricultural research: International investment patterns and policy perspectives. *World Development* 26 (6): 1057–71.

Alston, Julian M., and Raymond J. Venner. 2002. The effects of the U.S. Plant Variety Protection Act on wheat genetic improvement. *Research Policy* 31 (4): 527–42.

Arrow, Kenneth J., and Anthony C. Fisher. 1974. Environmental preservation, uncertainty, and irreversibility. *Quarterly Journal of Economics* 88 (2): 312–19.

Atkinson, Richard C., Roger N. Beachy, Gordon Conway, France A. Cordova, Marye Anne Fox, Karen A. Holbrook, Daniel F. Klessig, et al. 2003. Intellectual property rights: Public sector collaboration for agricultural IP management. *Science* 301 (5630): 174–75.

Barton, J. 1998. The impact of contemporary patent law on plant biotechnology research. In *Intellectual property rights III, global genetic resources: Access and property rights,* ed. S. A. Eberhardt, H. L. Shands, W. Collins, and R. L. Lower, 85–87. Madison, WI: Crop Sciences Society of America.

BBC News. 2000. Ministers "misled" public on BSE. *BBC News,* October 26. http://news.bbc.co.uk/2/hi/uk_news/992020.stm.

Berwald, Derek, Colin A. Carter, and Guillaume P. Gruère. 2006. Rejecting new technology: The case of genetically modified wheat. *American Journal of Agricultural Economics* 88 (2): 432–47.

Bessen, James, and Robert M. Hunt. 2003. An empirical look at software patents. Federal Reserve Bank of Philadelphia Working Paper no. 03-17. http://www.researchoninnovation.org/swpat.pdf.

Binenbaum, Eran, Carol Nottenburg, Philip G. Pardey, Brian D. Wright, and Patricia Zambrano. 2003. South-North trade, intellectual property jurisdictions, and freedom to operate in agricultural research on staple crops. *Economic Development and Cultural Change* 51 (2): 309–36.

Binswanger, Hans P., and Vernon W. Ruttan. 1978. *Induced innovation: Technology, institutions and development.* Baltimore, MD: Johns Hopkins University Press.

Boettiger, Sara, Gregory D. Graff, Philip G. Pardey, Eric Van Dusen, and Brian D. Wright. 2004. Intellectual property rights for plant biotechnology: International aspects. In *Handbook of plant biotechnology,* ed. Paul Christou and Harry Klee, 1089–1113. Chichester, UK: Wiley.

Boettiger, Sara, and Brian D. Wright. 2006. Open source in biotechnology: Open questions. *Innovations: Technology, Governance, and Globalization* 1 (4): 45–57.

Boldrin, Michele, and David K. Levine. 2008. *Against intellectual monopoly.* New York: Cambridge University Press. http://www.dklevine.com/general/intellectual /againstfinal.htm.

Brennan, Margaret, Carl Pray, Anwar Naseem, and James F. Oehmke. 2005. An innovation market approach to analyzing impacts of mergers and acquisitions in the plant biotechnology industry. *AgBioForum: The Journal of Agrobiotechnology Management and Economics* 8 (2–3): 89–99.

Brock, William H. 1997. *Justus von Liebig: The chemical gatekeeper.* Cambridge, UK: Cambridge University Press.

Broothaerts, Wim, Heidi J. Mitchell, Brian Weir, Sarah Kaines, Leon M. A. Smith, Wei Yang, Jorge E. Mayer, Carolina Roa-Rodriguez, and Richard A. Jefferson. 2005. Gene transfer to plants by diverse species of bacteria. *Nature* 433 (7026): 629–33.

Brush, S. B., J. E. Taylor, and M. R. Bellon. 1992. Technology adoption and biological diversity in Andean potato agriculture. *Journal of Development Economics* 39:365–87.

Busch, Lawrence, Richard Allison, Craig Harris, Alan Rudy, Bradley T. Shaw, Toby Ten Eyck, Dawn Coppin, Jason Konefal, Christopher Oliver, and James Fairweather. 2004. *External review of the collaborative research agreement between Novartis Agricultural Discovery Institute, Inc. and the Regents of the University of California.* East Lansing, MI: Institute for Food and Agricultural Standards (IFAS), Michigan State University.

Byerlee, Derek, and Ken Fischer. 2002. Accessing modern science: Policy and institutional options for agricultural biotechnology in developing countries. *World Development* 30 (6): 931–48.

Caswell, Margriet F., and David Zilberman. 1986. The effects of well depth and land quality on the choice of irrigation technology. *American Journal of Agricultural Economics* 68 (4): 798–811.

CNN. 2001. French police in BSE government raid. *CNN News,* January 18. http:// premium.asia.cnn.com/2001/WORLD/europe/france/01/17/france.madcow .02/index.html.

Consultative Group on International Agricultural Research (CGIAR). 2008. The origins of the CGIAR. Washington, DC: Consultative Group on International Agricultural Research. http://www.cgiar.org/who/history/origins.html.

Consultative Group on International Agricultural Research (CGIAR) Independent Review Panel. 2008. *Bringing together the best of science and the best of development.* Independent review of the CGIAR system, report to the executive council. Washington, DC: Consultative Group on International Agricultural Research. http://www.cgiar.org/externalreview/.

Cohen, Wesley M., Richard R. Nelson, and John P. Walsh. 2000. Protecting their intellectual assets: Appropriability conditions and why U.S. manufacturing firms patent (or not). NBER Working Paper no. 7552. Cambridge, MA: National Bureau of Economic Research.

Cohen, Joel I., and Robert Paarlberg. 2004. Unlocking crop biotechnology in developing countries—A report from the field. *World Development* 32 (9): 1563–77.

Constantine, John H., Julian M. Alston, and Vincent H. Smith. 1994. Economic impacts of the California One-Variety Cotton Law. *Journal of Political Economy* 102 (5): 951–74.

Coughenour, C. M., and Shankariah Chamala. 2001. *Conservation tillage and cropping innovation: Constructing the new culture of agriculture.* Hoboken, NJ: Wiley-Blackwell.

Daley, Suzanne. 2001. French report faults response to mad cow crisis. *New York Times,* May 18. http://www.nytimes.com/2001/05/18/world/french-report-faults -response-to-mad-cow-crisis.html.

David, Paul A. 1969. *A contribution to the theory of diffusion.* Stanford Center for Research in Economic Growth Memorandum no. 71. Stanford, CA: Stanford University.

Day-Rubenstein, Kelly, and Keith O. Fuglie. 2000. The CRADA model for public-private research and technology transfer in agriculture. In *Public-private collaboration in agricultural research: New institutional arrangements and economic implications,* ed. Keith O. Fuglie and David E. Schimmelpfennig, 155–174. Ames, IA: Iowa State University Press.

Delmer, Deborah P., Carol Nottenburg, Greg D. Graff, and Alan B. Bennett. 2003. Intellectual property resources for international development in agriculture. *Plant Physiology* 133 (4): 1666–70.

Erbisch, F. H. 2000. Challenges of plant protection: How a semi-public agricultural research institution protects its new plant varieties and markets them. Paper presented at workshop, The Impact on Research and Development of Sui Generis Approaches to Plant Variety Protection of Rice in Developing Countries, Los Baños, The Philippines.

Evenson, Robert E. 1980. A century of agricultural research and productivity change research, invention, extension, and productivity change in U.S. agriculture: An historical decomposition analysis. In *Research and extension productivity in agriculture,* ed. A. A. Araji, 146–228. Moscow, ID: Department of Agricultural Economics, University of Idaho.

———. Economic impacts of agricultural research and extension. In *Handbook of agricultural economics.* Vol. 1A, Agricultural Production, ed. Bruce L. Gardner and Gordon C. Rausser, 573–628. New York: Elsevier Science/North Holland.

Falck-Zepeda, José Benjamin, Greg Traxler, and Robert G. Nelson. 2000. Surplus distribution from the introduction of a biotechnology innovation. *American Journal of Agricultural Economics* 82 (2): 360–69.

Falck-Zepeda, Jose, and Greg Traxler. 2000. The role of federal, state, and private institutions in seed technology generation. In *Public-private collaboration in agricultural research: New institutional arrangements and economic implications,* ed. Keith O. Fuglie and David E. Schimmelpfennig, 99–115. Ames, IA: Iowa State University Press.

Feder, Gershon. 1980. Farm size, risk aversion and the adoption of new technology under uncertainty. *Oxford Economic Papers* 32 (2): 263–83.

———. 1982. Adoption of interrelated agricultural innovations: Complementarity and impact of risk, scale, and credit. *American Journal of Agricultural Economics* 64 (1): 94–101.

Feder, Gershon, Richard E. Just, and David Zilberman. 1985. Adoption of agricultural innovations in developing countries: A survey. *Economic Development and Cultural Change* 33 (2): 255–98.

Fernandez-Cornejo, Jorge, E. Douglas Beach, and Wen-Yuan Huang. 1994. The adoption of IPM techniques by vegetable growers in Florida, Michigan, and Texas. *Journal of Agricultural and Applied Economics* 26 (1): 158–72.

Fernandez-Cornejo, Jorge, Chad Hendricks, and Ashok Mishra. 2005. Technology adoption and off-farm household income: The case of herbicide-tolerant soybeans. *Journal of Agricultural and Applied Economics* 37 (3): 549–63.

Finlay, Mark R. 1988. The German agricultural experiment stations and the beginnings of American agricultural research. *Agricultural History* 62 (2): 41–50.

Foltz, Jeremy D., Bradford L. Barham, and Kwansoo Kim. 2007. Synergies or tradeoffs in university life sciences research. *American Journal of Agricultural Economics* 89 (2): 353–67.

Fowler, Cary. 1994. *Unnatural selection: Technology, politics, and plant evolution.* Yverdon, Switzerland: Gordon and Breach.

Friedman, David D., William M. Landes, and Richard A. Posner. 1991. Some economics of trade secret law. *Journal of Economic Perspectives* 5 (1): 61–72.

Fuglie, Keith, Nicole Ballenger, Kelly Day, Cassandra Klotz, Michael Ollinger, John Reilly, Utpal Vasavada, and Jet Yee. 1996. *Agricultural research and development: Public and private investments under alternative markets and institutions.* Agricultural Economic Report Number no. 735. Washington, DC: Economic Research Service, U.S. Department of Agriculture.

Glenna, Leland L., William B. Lacy, Rick Welsh, and Dina Biscotti. 2007. University administrators, agricultural biotechnology and academic capitalism: Defining the public good to promote university-industry relationships. *Sociological Quarterly* 48 (1): 141–63.

Gollier, Christian, B. Moldovanu, and T. Ellingsen. 2001. Should we beware of the precautionary principle? *Economic Policy* 16 (33): 303–27.

Graff, Gregory D., Susan E. Cullen, Kent J. Bradford, David Zilberman, and Alan B. Bennett. 2003. The public-private structure of intellectual property ownership in agricultural biotechnology. *Nature Biotechnology* 21 (9): 989–95.

Graff, Gregory D., Brian D. Wright, Alan B. Bennett, and David Zilberman. 2004. Access to intellectual property is a major obstacle to developing transgenic horticulture crops. *California Agriculture* 58 (2): 120–26.

Graff, Gregory D., and David Zilberman. 2007. The political economy of intellectual property: Reexamining European policy on plant biotechnology. In *Agricultural biotechnology and intellectual property: Seeds of change,* ed. Jay Kesan, 244–267. Washington, DC: CAB International.

Griliches, Zvi. 1957. Hybrid corn: An exploration in the economics of technological change. *Econometrica* 25 (4): 501–22.

Guttman, Joel M. 1978. Interest groups and the demand for agricultural research. *Journal of Political Economy* 86 (3): 467–84.

Harding, T. Swann. 1940. Henry L. Ellsworth, Commissioner of Patents. *Journal of Farm Economics* 22 (3): 621–27.

Hayami, Yujiro, and Vernon W. Ruttan. 1971. *Agricultural development.* Baltimore, MD: Johns Hopkins University Press.

Huang, Jikun, Ruifa Hu, Scott Rozelle, and Carl Pray. 2005. *Development, policy, and impacts of genetically modified crops in China: A comprehensive review of China's agricultural biotechnology sector.* Belfer Center for Science and International Affairs, John F. Kennedy School of Government, Harvard University, Harvard University, Working Paper.

Huang, Jikun, Scott Rozelle, Carl Pray, and Qinfang Wang. 2002. Plant biotechnology in China. *Science* 295 (5555): 674–76.

Huffman, Wallace E., and Robert E. Evenson. 2006. *Science for agriculture: A long-term perspective.* Ames, IA: Iowa State University Press.

Huffman, Wallace E., and John A. Miranowski. 1981. An economic analysis of expenditures on agricultural experiment station research. *American Journal of Agricultural Economics* 63 (1): 104–18.

International Maize and Wheat Improvement Center (CIMMYT). 1997. *CIMMYT in 1995–1996: The next 30 years.* Mexico, D. F.: International Maize and Wheat Improvement Center.

James, C. 2008. *Global status of commercialized biotech/GM crops: 2008.* ISAAA Brief no. 39. Ithaca, NY: International Service for the Acquisition of Agri-Biotech Applications.

Jefferson, Richard. 2006. Science as social enterprise: The Cambia BiOS Initiative. *Innovations: Technology, Governance, and Globalization* 1 (4): 13–44.

Joly, Pierre-Benoit, and Marie-Angèle de Looze. 1996. An analysis of innovation strategies and industrial differentiation through patent applications: The case of plant biotechnology. *Research Policy* 25 (7): 1027–46.

Judd, M. Ann, James K. Boyce, and Robert E. Evenson. 1986. Investing in agricultural supply: The determinants of agricultural research and extension investment. *Economic Development and Cultural Change* 35 (1): 77–113.

———. 1987. Investment in agricultural research and extension. In *Policy for agricultural research,* ed. Vernon W. Ruttan and Carl E. Pray, 7–38. Boulder, CO: Westview Press.

Juma, Calestous. 1989. *The gene hunters: Biotechnology and the scramble for seeds.* Princeton, NJ: Princeton University Press.

Just, Richard E., and Wallace E. Huffman. 1992. Economic principles and incentives: Structure, management, and funding of agricultural research in the United States. *American Journal of Agricultural Economics* 74 (5): 1101–8.

———. 2009. The economics of universities in a new age of funding options. *Research Policy* 38 (7): 1102–16.

Just, Richard E., and Gordon C. Rausser. 1993. The governance structure of agriculture science and agricultural economics: A call to arms. *American Journal of Agricultural Economics,* 75 (special issue): 69–83.

King, John L., and David Schimmelpfennig. 2005. Mergers, acquisitions, and stocks of agricultural biotechnology intellectual property. *AgBioForum: The Journal of Agrobiotechnology Management and Economics* 8 (2–3): 83–88.

Knowler, Duncan, and Ben Bradshaw. 2007. Farmers' adoption of conservation agriculture: A review and synthesis of recent research. *Food Policy* 32 (1): 25–48.

Knudson, Mary K. 2000. The research consortium model for agricultural research. In *Public-private collaboration in agricultural research: New institutional arrangements and economic implications,* ed. Keith O. Fuglie and David E. Schimmelpfennig, 175–98. Ames, IA: Iowa State University Press.

Knudson, Mary K., and Carl E. Pray. 1991. Plant variety protection, private funding, and public sector research priorities. *American Journal of Agricultural Economics* 73 (3): 882–86.

Koo, Bonwoo, and Brian D. Wright. 1999. The role of biodiversity products as incentives for conserving biological diversity: Some instructive examples. *Science of the Total Environment* 240 (1–3): 21–30.

Lei, Zhen, Rakhi Juneja, and Brian D. Wright. 2009. Patents versus patenting: Implications of intellectual property protection for biological research. *Nature Biotechnology* 27 (1): 36–40.

Lele, Uma. 2003. Biotechnology: Opportunities and challenges for developing countries. *American Journal of Agricultural Economics* 85 (5): 1119–25.

Lele, U., C. Barrett, C. K. Eicher, B. Garnde, C. Gerrard, L. Kelly, W. Lesser, K. Perkins, S. Rana, M. Rukuni, and D. J. Speilman. 2003. *The CGIAR at 31: An independent meta-evaluation of the Consultative Group on International Agricultural Research.* Operations Evaluation Department. Washington, DC: World Bank.

Lemarié, Stéphane. 2001. How will U.S.-based companies make it in Europe? An insight from Pioneer and Monsanto. *AgBioForum: The Journal of Agrobiotechnology Management and Economics* 4 (1): 74–78.

Lemley, Mark A., and Carl Shapiro. 2005. Probabilistic patents. *Journal of Economic Perspectives* 19 (2): 75–98.

———. 2007. Patent hold-up and royalty stacking. *Texas Law Review* 85 (7): 1991–2049.

Levin, Richard C., Alvin K. Klevorick, Richard R. Nelson, Sidney G. Winter, R. Gilbert, and Z. Griliches. 1987. Appropriating the returns from industrial research and development. *Brookings Papers on Economic Activity* 18 (1987-3): 783–831.

Lusk, Jayson L., Jutta Roosen, and John A. Fox. 2003. Demand for beef from cattle administered growth hormone or fed genetically modified corn: A comparison of consumers in France, Germany, the United Kingdom, and the United States. *American Journal of Agricultural Economics* 85 (1): 16–29.

Marco, Alan C., and Gordon C. Rausser. 2008 The role of patent rights in mergers: Consolidation in plant biotechnology. *American Journal of Agricultural Economics* 90 (1): 133–51.

Matuschke, Ira, Ritesh R. Mishra, and Matim Qaim. 2007. Adoption and impact of hybrid wheat in India. *World Development* 35 (8): 1422–35.

Miller, Henry I., and Gregory Conko. 2000. The science of biotechnology meets the politics of global regulation. *Issues in Science and Technology* 17 (Fall): 47–54.

Murray, Fiona, and Scott Stern. 2007. Do formal intellectual property rights hinder the freedom of scientific knowledge? An empirical test of the anti-commons hypothesis. *Journal of Economic Behavior and Organization* 63 (4): 648–87.

National Research Council. 1995. *Colleges of agriculture at the land grant universities: A profile.* Washington, DC: National Academies Press.

Nelson, Carl H. 2001. Risk perception, behavior, and consumer response to genetically modified organisms: Toward understanding American and European public reaction. *American Behavioral Scientist* 44 (8): 1371–88.

Oehmke, James F., and Christopher A. Wolf. 2004. Is Monsanto leaving money on the table? Monopoly pricing and Bt cotton value with heterogeneous adopters. *Journal of Agricultural and Applied Economics* 36 (3): 705–18.

O'Hanlon, Kevin. 2004. Starlink corn settlement also to include interest. *USA Today,* August 23.

Ollinger, Michael, and Jorge Fernandez-Cornejo. 1995. Regulation, innovation, and market structure in the U.S. pesticide industry. AER-719. Washington, DC: Economic Research Services, U.S. Department of Agriculture.

Olmstead, Alan L., and Paul Rhode. 1993. Induced innovation in American agriculture: A reconsideration. *Journal of Political Economy* 101 (1): 100–118.

Pardey, Philip G., Julian M. Alston, and Roley R. Piggott, eds. 2006. *Agricultural R&D in the developing world: Too little, too late?* Washington, DC: International Food Policy Research Institute.

Pardey, Philip G., and Nienke M. Beintema. 2001. *Slow magic: Agricultural R&D a century after Mendel.* IFPRI Food Policy Report. Washington, DC: International Food Policy Research Institute.

Pardey, Philip G., Nienke M. Beintema, Steven Dehmer, and Stanley Wood. 2006. *Agricultural research: A growing global divide?* IFPRI Food Policy Report, Agricultural Science and Technology Indicators Initiative. Washington, DC: International Food Policy Research Institute.

Pardey, Philip G., Jennifer James, Julian Alston, Stanley Wood, Bonwoo Koo, Eran Binenbaum, Terrence Hurley, and Paul Glewwe. 2007. *Science, technology, and skills.* International Science and Technology Practice and Policy (INSTEPP). Rome: CGIAR and Department of Applied Economics, University of Minnesota, Food and Agriculture Organization of the United Nations.

Park, Chris C. 1989. *Chernobyl: The long shadow.* London: Routledge.

Parker, Douglas D., and David Zilberman. 1993. University technology transfers: Impacts on local and U.S. economies. *Contemporary Policy Issues* 11 (2): 87–100.

Pollack, A. 2001. Altered corn surfaced earlier. *New York Times,* September 4.

Polson, Rudolph A., and Dunstan S. C. Spencer. 1991. The technology adoption process in subsistence agriculture: The case of cassave in southwestern Nigeria. *Agricultural Systems* 36 (1): 65–78.

Poortinga, Wouter, and Nick F. Pidgeon. 2005. Trust in risk regulation: Cause or consequence of the acceptability of GM food? *Risk Analysis* 25 (1): 199–209.

Pray, Carl E., Prajakta Bengali, and Bharat Ramaswami. 2005. The cost of biosafety regulations: The Indian experience. *Quarterly Journal of International Agriculture* 44 (3): 267–89.

Pray, Carl E., and Keith O. Fuglie. 2001. *Private investment in agricultural research and international technology transfer in Asia.* Economic Research Service Technical Report no. 805. Washington, DC: U.S. Department of Agriculture.

Qaim, Matim, Arjunan Subramanian, Gopal Naik, and David Zilberman. 2006. Adoption of Bt cotton and impact variability: Insights from India. *Review of Agricultural Economics* 28 (1): 48–58.

Rahm, Michael R., and Wallace E. Huffman. 1984. The adoption of reduced tillage: The role of human capital and other variables. *American Journal of Agricultural Economics* 66 (4): 405–13.

Rausser, Gordon. 1999. Private/public research: Knowledge assets and future scenarios. *American Journal of Agricultural Economics* 81 (5): 1011–27.

Rausser, Gordon, Leo Simon, and Holly Ameden. 2000. Public-private alliances in biotechnology: Can they narrow the knowledge gaps between rich and poor? *Food Policy* 25 (4): 499–513.

Rose-Ackerman, Susan, and Robert Evenson. 1985. The political economy of agricultural research and extension: Grants, votes, and reapportionment. *American Journal of Agricultural Economics* 67 (1): 1–14.

Runge, C. Ford, and Barry Ryan. 2004. The global diffusion of plant biotechnology: International adoption and research in 2004. Department of Applied Economics, University of Minnesota, St. Paul. Unpublished Report.

Ruttan, Vernon W. 1977. The Green Revolution: Seven generalizations. *International Development Review* 19:16–23.

———. 1980. Bureaucratic productivity: The case of agricultural research. *Public Choice* 35 (5): 529–49.

———. 1982. *Agricultural research policy.* Minneapolis, MN: University of Minnesota Press.

Schoenermarck, H. 1974. Nathaniel Bagshaw Ward: How a sphinx moth altered the ecology of earth. *Garden Journal* 24:148–54.

Shuren, Jeffrey. 2008. Guidance for industry on the Food and Drug Administration: Recommendations for sampling and testing yellow corn and dry-milled yellow corn shipments intended for human food use for Cry9C protein residues; withdrawal of guidance. Docket no. FDA-2008-D-0229. Washington, DC: U.S. Food and Drug Administration, Department of Health and Human Services.

Sjöberg, Lennart. 2001. Limits of knowledge and the limited importance of trust. *Risk Analysis* 21 (1): 189–98.

Skinner, Jonathan S., and Douglas Staiger. 2005. Technology adoption from hybrid corn to beta blockers. NBER Working Paper Series w11251. Available at SSRN: http://ssrn.com/abstract=697176.

Stokes, Donald E. 1997. *Pasteur's quadrant: Basic science and technological innovation.* Washington, DC: Brookings Institution.

Sunding, David, and David Zilberman. 2001. The agricultural innovation process: Research and technology adoption in a changing agricultural sector. In *Handbook of agricultural economics.* Vol. 1A, ed. Bruce L. Gardner and Gordon C. Rausser, 201–61. Amsterdam, The Netherlands: Elsevier B.V.

Taylor, Michael R., and Jerry Cayford. 2002. The U.S. patent system and developing country access to biotechnology: Does the balance need adjusting? Resources for the Future Discussion Paper. Washington, DC: Resources for the Future.

Thomson, Jennifer A. 2004. The status of plant biotechnology in Africa. *AgBioForum* 7 (1 and 2): 9–12.

Tollens, Eric, Matty Demont, and Rony Swennen. 2004. Agrobiotechnology in developing countries: North-South partnerships are the key. *Outlook on Agriculture* 33 (4): 231–38.

Troyer, A. Forrest. 2009. Development of hybrid corn and the seed corn industry. In *Handbook of maize—Volume II: Genetics and genomics,* ed. Jeffrey L. Bennetzen and Sarah Hake, 87–114. New York: Springer Science+Business Media.

Ulrich, A., Hartley Furtan, and Andrew Schmitz. 1986. Public and private returns from joint venture research: An example from agriculture. *Quarterly Journal of Economics* 88 (2): 312–19.

United States Department of Justice. 1998. Justice Department approves Monsanto's acquisition of DeKalb Genetics Corporation. Press Release no. 98-570. Washington, DC: U.S. Department of Justice. http://www.usdoj.gov/atr/public/press_releases/1998/2103.htm.

Vogel, David. 2001. *The new politics of risk regulation in Europe.* London: Centre for Analysis of Risk and Regulation, London School of Economics and Political Science.

Vogt, Donna U., and Mickey Parish. 1999. Food biotechnology in the United States: Science, regulation, and issues. Congressional Research Service Reports and Issue Briefs. Washington, DC: Domestic Social Policy Division, U.S. Department of State.

Weingart, Peter. 2002. The moment of truth for science. *EMBO Reports* 3 (8): 703–06.

White, Fred C., and Joseph Havlicek, Jr. 1982. Optimal expenditures for agricultural research and extension: Implications for underfunding. *American Journal of Agricultural Economics* 64 (1): 47–55.

World Bank Group. 2009. *Implementing agriculture for development.* The World Bank Group Agriculture Action Plan: FY2010-2012. Washington, DC: World Bank. http://siteresources.worldbank.org/INTARD/Resources/Agriculture_Action_Plan_web.pdf.

Wright, Brian D. 1998. Public germplasm development at a crossroads: Biotechnology and intellectual property. *California Agriculture* 52 (6): 8–13.

Wright, Brian D., and Philip G. Pardey. 2006. Changing intellectual property regimes:

Implications for developing country agriculture. *International Journal of Technology and Globalization* 2 (1-2): 93–114.

Wright, Brian D., and Philip G. Pardey, Carol Nottenburg, and Bonwoo Koo. 2007. Agricultural innovation: Investments and incentives. In *Handbook of agricultural economics.* Vol. 3, part 3, ed. Robert Evenson and Prabhu Pingali, 2535–2605. Amsterdam, The Netherlands: Elsevier B.V.

Zilberman, David. 2006. The economics of biotechnology regulation. In *Regulating agricultural biotechnology: Economics and policy,* ed. Richard E. Just, Julian M. Alston, and David Zilberman, 243–61. New York: Springer Science+Business Media.

3

Implications for Energy Innovation from the Chemical Industry

Ashish Arora and Alfonso Gambardella

Once a leader in industrial innovation, the chemical industry has changed countless aspects of modern life. From the plastic in the toothbrush we use in the morning, to the tires we drive to work on and the fuel that powers them, to the clothes that keep us warm, chemical innovations are so infused in our daily lives that we generally take them for granted.

It is difficult, therefore, to speak of *the* chemical industry, one that David Landes (1969, 269) has called the most "miscellaneous of industries" and which encompasses synthetic fibers, plastics, agricultural pesticides and fertilizers, food additives, health and beauty aids, and many other products and production components. Given this variety, attempting to summarize chemical innovation is difficult. Instead, we shall focus on industrial synthetic fibers and plastics and the inputs from which they are made. Their history offers useful lessons for how energy innovation and diffusion might be accelerated, in part because innovations in these chemical subsectors share common features with energy innovations.

Chemical innovation has been marked by the search for new inputs and the concomitant process innovations that allow the inputs to be produced and used. A prominent example of such a change is the shift from coal derived inputs for producing synthetic fibers and plastics to those from oil and natural gas. This change, which began before World War II in the United

Ashish Arora is a professor at the Fuqua School of Business, Duke University, and a research associate of the National Bureau of Economic Research. Alfonso Gambardella is a professor in the Department of Management at Bocconi University.

We are grateful to Linda Cohen, Rebecca Henderson, Richard Newell, and other participants at the National Bureau of Economic Research (NBER) meeting, April 2009, Boston, for helpful comments, and to Bonnie Nevel and Susan Schaffer for assistance in the preparation of this manuscript. We remain responsible for all remaining errors.

States, was accompanied by a flood of innovations, including advances in petroleum refining and new processes for the production of important synthetic fibers and plastics. Energy innovation, too, is likely to involve the development and large-scale deployment of new processes and new materials designed by chemists and chemical engineers.

Innovation in the chemical industry requires scaling-up laboratory discoveries into commercially viable products. This sort of scale-up is also common in the oil refining industry, and, indeed, chemical engineering has its roots in oil refining. The gigantic refineries that produce gasoline, kerosene, diesel, and a host of valuable inputs used in a wide range of commonly used materials embody numerous innovations in chemistry and chemical engineering. The biorefineries of the future will no doubt be different from the oil refineries of the present, but much of the expertise they require will come from chemists and chemical engineers.

There are more subtle connections as well. The successful commercial introduction of new materials has required careful attention to questions such as how downstream users would use them, whether existing processing machinery would have to be modified, and how the final product in which the new material was embodied would differ. All these considerations have effects that ripple through the value chain, which is typical of energy innovations as well.[1] For example, the introduction of a new automobile fuel may also require changes in how it is distributed and used, and a new form of rubber with different temperature-related properties may require new learning about best applications for its use. Chemical innovation, which has often consisted of new types of materials that are used to produce existing products, have long featured this type of adjustment.

Despite this complexity, for the most part, the chemical industry has relied upon markets to coordinate the required changes. This coordination is far from perfect. Many chemical processes have produced harmful by-products. However, the history of chemical innovation also reflects the search to mitigate or eliminate waste. For instance, chlorine produced as a by-product in producing caustic soda in the mid-nineteenth century was used to produce bleaching powder. Changes in the market place, especially growth in demand, have occasionally made what was hitherto waste more valuable and, therefore, worth capturing. Thus, in the nineteenth century in the United States, kerosene was the principal product produced from crude oil. Natural gas was routinely flared in oil pumps, and heavier components in crude oil, such as waxes, were simply tossed aside. Over time, in response to rising demand, techniques were developed to use these hitherto waste products. The implications for energy innovation, which centrally involves the problem of dealing with harmful by-products such as carbon dioxide, are obvious.

---

1. A "value chain" is a series of activities for which each step of the activity increases the value of the target product.

Major innovations in the recent history of the chemical industry have been the result of privately funded research carried out in the laboratories of large chemical and oil firms. In the 1870s, when the chemical industry revolved around synthetic dyes, firms believed that hiring chemists would help them discover new dyes, the way to commercial success. During the 1920s and years of the Great Depression, the spread of the automobile created demand for gasoline (and, hence, for new refining technologies), lacquers (as car paint), plastics, and materials for car tires. Later, World War II greatly boosted the demand for plastics, synthetic fibers such as nylon and polyester, high octane gasoline, and synthetic rubber. In each case, the payoffs for developing new materials, improving them, and developing techniques to reduce production costs and increase production volumes were significant and predictable. The post–World War II boom in the 1950s and 1960s further increased demand for synthetic fibers such as polyester and for new plastics such as polystyrene.

This commitment to investing in research, not simply in production, became ingrained in the industry. Hounshell and Smith (1988) argue that the commercial success of nylon, commercialized just as World War II started, may have been salient in shaping the thinking of the management of DuPont, and by extension, of other chemical firms as well. Hounshell and Smith describe in detail the largely fruitless attempts of DuPont management to search for blockbuster products that would be as successful as nylon. In the process, the company built substantial in-house capability for research and development (R&D), and for some time even operated an in-house "Polymer Institute." By contrast, most other technology-intensive sectors developed after World War II, by which time government support for research was more forthcoming and, in some cases, decisive in starting the industry.

Early in the history of the industry, when synthetic dyestuffs were based on advances in organic chemistry by German chemists, universities made formative contributions, and in the twentieth century, they have contributed indirectly (but importantly) by institutionalizing the learning being created by firms and by training students—creating the disciplines, as it were, of petroleum engineering, chemical engineering, polymer chemistry, polymer engineering, and so on. The federal government's role in supporting innovation in the industry has not, however, been limited to supporting universities and programs such as synfuels and the synthetic rubber program. It has also played a key role in facilitating the development of a "market for technology" in the industry, through both antitrust and intellectual property policy. But it has been in private research labs, and indeed through the research programs of a wide range of private firms including smaller players and specialized engineering firms, or SEFs, that most recent discoveries have taken place.

The development of new processes and materials alone does not make a significant contribution to economic growth; the new processes and materials have to be widely diffused. In the chemical industry, this diffusion was largely market based. Direct government subsidies were mostly unimportant. The rapid diffusion of synthetic fibers and plastics, such as nylon, polyester, and polystyrene was fostered by an extensive market for technology in which chemical technologies were widely licensed. Specialized engineering firms, which supplied technology and know-how, played a key role in this process. The emergence of these markets for technology appears to have been importantly due to antitrust policies, principally in the United States, and the enforcement of intellectual property rights.

In this chapter, we begin by laying out the early history of the chemical industry for an overview of the role innovation has played in its development. We then explore three noteworthy historical experiences. We describe the switch in coal-based raw materials to those derived from oil and gas and briefly analyze two government programs that have attempted to promote innovation: synthetic rubber and synfuels. We take a close look at the role that specialized engineering firms have played in the diffusion of important innovations, and we detail the effect that government policies have had on fostering innovation.

3.1 Early History

The modern chemical industry began with organic chemicals and, specifically, synthetic dyes. Beginning with William Perkin's accidental discovery of a mauve (purple) dye in 1856 while in Professor August Hofmann's lab, new synthetic dyes were rapidly and subsequently discovered in France and Germany. By the 1870s, German firms dominated the synthetic dye market. Although the initial discoveries were made in the labs of university professors—there were few chemical labs outside universities—private firms quickly began investing in their own labs: by the 1880s, the leading German firms had created in-house laboratories for discovering new synthetic dyes. Ten years later, the vast majority of synthetic dyes came from the R&D labs of German chemical companies.

University research remained important for opening up fields of inquiry and of suggesting fruitful areas of investigation. For instance, Friedrich August Kekulé's discovery of the benzene ring structure in 1866 was crucial for the discovery of dyes based on aniline by enabling researchers to predict what colors different reagents might produce. By the 1880s, university research was aimed at clarifying the structure of dyes discovered in corporate labs, and understanding their properties, rather than discovering new dyes as such (Murmann 2003).

The United States was a follower in scientific chemical research and also a net technology importer until World War I. During this time, American chemical firms focused on developing new ways of producing chemicals on a large scale, especially commodity chemicals like sulfuric acid (Arora and

Table 3.1   **Major polymer innovations: inventors and commercializers**

Polymer	Inventor	Organization of inventor	Commercializing firm	Year of commercialization
		Synthetic fibers		
Nylon	Carothers	DuPont	DuPont	1934
Polyester	Whinfield-Dickson	Calico Printers	ICI	1939
Acrylics	Various		Bayer; DuPont	1920–1930s
		Plastics		
Phenolic resins	Baekland; Edison and others		General Bakelite	1910
PVC	Fritz Klatte	Griesham Electron	Union Carbide; BF Goodrich and GE	1930s
Polystyrene	Various		IG Farben	1930
Low density polyethylene	Swallow, Perrin, Fawcett, Gibson	ICI	ICI	1939
Polypropylene	Ziegler-Natta; Hogan & Banks	Max-Planck Inst.; Montecatini; Phillips Petroleum	Various	1960s
Neoprene	Nieuwland, Collins	Notre Dame University; DuPont	DuPont	1936

Source: Arora (2002).

Rosenberg 1998). These innovations were based in experience, rather than scientific discovery—that is, the innovations arose from trial and error in the lab, rather than relying on recent scientific advances. Products such as fertilizers and gasoline also required large-scale production to reduce costs. As petroleum refining grew in importance, so did the demand for new refining processes.

The period from 1920 to 1960 probably marks the golden age of innovation in the chemical industry, at least as far as the United States is concerned. The major polymers—plastics and synthetic fibers—originated in corporate labs and were commercialized during this forty-year period (table 3.1).[2] This period also marks major developments in chemical engineering, with the commercialization of a number of important chemical processes. In many cases, fundamental scientific contributions came from researchers working for private firms, such as Wallace Carothers (DuPont), Frank Mayo (U.S. Rubber), and Giulio Natta (Montecatini). Paul Flory, later to receive the Nobel Prize in chemistry for his contributions to the area of polymer chemistry, worked for many years in the research departments of DuPont, Standard Oil, and other chemical firms before moving to academia.

2. The two exceptions are the first synthetic rubber (neoprene), which was based on work by Julius Nieuwland at Notre Dame University, and the catalyst systems used to produce polypropylene and high density polyethylene, developed by Karl Ziegler at what is now part of the Max Planck Institutes.

Chemical engineering was a distinctly American achievement—more specifically, it was a testimony to America's very productive university-industry interface.[3] The signal contribution of universities was to train students and institutionalize the disciplines of chemical engineering, petroleum engineering, polymer chemistry, and polymer engineering (Rosenberg 1998). Many of the basic technological breakthroughs came from researchers trained in these new disciplines and moving on to corporate labs. For example, the first significant chemical process innovation, the Haber-Bosch process, was developed by BASF early in the twentieth century. The general-purpose nature of chemical engineering enabled university research and training to play an important role in applying engineering science to the practical problem of designing large-scale processes. A number of the major advances in catalytic refining techniques were developed by oil firms, notably Standard Oil of New Jersey (now ExxonMobil). University researchers were sometimes involved in these innovations, but typically in partnership with researchers from corporate labs. Notably, Warren Lewis and Edwin Gilliland from Massachusetts Institute of Technology (MIT) developed fluidized-bed catalytic cracking in close cooperation with the chemical engineers at Standard Oil (Spitz 1988).

Individuals and small engineering firms (which were typically entrepreneurial start-up firms) were another important source of chemical process innovation during the golden age. For instance, Scientific Design, a small firm founded in 1946, developed new fixed-bed catalytic processes for ethylene glycol, maleic anhydride, and, most important, for production of purified terephthalic acid, the basic building block of polyester. Similarly, National Hydrocarbon Company (now Universal Oil Products [UOP]), provided some of the fundamental advances in refining technology, including thermal catalytic cracking, reforming, and sulfur extraction. Universal Oil Products remains a leading source of refining technology to this day.

3.2   Case Studies in Chemical Innovation

New materials used by firms as inputs (rather than sold directly to consumers) typically take a long time to diffuse broadly. Not only are industrial firms inherently conservative in their adoption of new materials, diffusion also takes time because new materials often require complementary changes

---

3. See Landau and Rosenberg (1992) for a discussion of the role of MIT in the development of chemical engineering as a discipline. The large size of the market had introduced American firms to the problems involved in scaling-up production of basic products, such as chlorine, caustic soda, soda ash, and sulfuric acid as early as the beginning of the twentieth century. This focus on large-scale production had additional benefits when it turned out that the new petrochemical technologies had strong plant-level economies of scale. Because scaling-up output was not a simple matter—it involved considerable learning—early experience with process technologies gained American firms a head start when petrochemicals became the dominant feedstock after World War II.

in the physical infrastructure. This issue is likely to arise in the case of energy. Fuel cells for automobiles are a case in point. The concomitant changes needed in the physical infrastructure (such as fueling stations) are thought to involve large financial investments and coordination problems, greatly delaying the widespread adoption of fuel cells (see, for instance, Struben and Sterman 2008).

Another significant cause of delay is what Bresnahan and Greenstein (1996) dub "co-invention" by users—that is, the need for users of the new technology to become familiar with its use and in some cases to design complementary adaptations to facilitate its use. Bresnahan and Greenstein found that co-invention can pose significant costs in time and money to a company, as demonstrated by the resistance and delay in the switch from mainframes to client-servers in U.S. firms. David (1990) similarly argues that the diffusion of electrical power was slowed as manufacturers, used to steam power, had to learn how to exploit the full potential of this new energy source. In the case of the chemical industry, Hounshell (1988) shows that as new industrial materials were introduced, users had to learn about the properties of these new materials. For instance, when first introduced, rayon fiber was believed to be weaker than cotton fiber, and because rayon fibers were further weakened when wet, the material was assumed to be unsuitable for tire cord. However, it was discovered that hot rayon is 50 percent stronger than cotton. Rayon tire cord, however, did not take off until World War II, when circumstances forced the use of synthetic rubber for tires; synthetic rubber tires run hotter than natural rubber tires, and, therefore, rayon was a better cord material under those hotter conditions. This also illustrates the systemic interdependencies in chemical innovation.

However, in general, diffusion in the chemical industry has been surprisingly fast. The most notable is the swift and thorough switch from coal to petroleum as the dominant chemical feedstock after World War II.

3.2.1   Coal to Petroleum

At the end of World War I, coal was central to the chemical industry, not so much as an energy source but as a source of raw materials: coal-derived chemicals formed the basis of the chemical industry. For instance, coal tar provided a source for synthetic dyes, coal coking furnaces provided the nitrogen for fertilizers and explosives such as TNT, and coal provided the toluene for those explosives. Heating coal to make coke—a raw material in the manufacture of high-carbon steel—also produces a number of chemically useful gases and byproducts: coal tar, ammonia, and benzene. Indeed, firms such as Solvay, Koppers, and Allied Chemicals owned batteries of by-product coke ovens to produce these valuable chemicals. These joined a large complementary physical infrastructure of by-product coke ovens across the United States and around Europe, complementing a substantial accumulated learning about the use of coal by-products as chemical building blocks. As late as 1938, DuPont, the leading American chemical company, announced that it would create a synthetic fiber "from coal, air, and water" when starting its first commercial nylon plant in Seaford, Delaware.

Everything changed with the meteoric postwar popularity of the automobile, which greatly increased the demand for gasoline, leading refiners to improve oil-refining processes. Furthermore, the oil-refining process for automobile fuel used only a portion of the petroleum refined, leaving ample by-product with no clear use. During the 1940s, petroleum by-products from automobile oil refineries increasingly went into the production of basic materials, such as plastics, synthetic rubber, and synthetic fibers. Oil-based chemicals began to quickly outstrip the previously dominant coal-based chemicals.

The change had many dimensions. New extracting and refining technology for petroleum had to be developed. Further, there was a concomitant change in the technology from acetylene-based chemistry (triple bonds typical of coal by-products) to the ethylene-based chemistry (double bonds) of oil by-products. But, ultimately, the move from coal to oil was driven by the superiority of oil-based feedstocks and by falling costs of oil due to the progressive improvements in extraction and refining technologies.

The switch from coal to oil also is remarkable because of its rapid worldwide diffusion. That these processes were rapidly adopted in Europe is significant because Europe was abundantly supplied with coal (but little oil) and because Europe had pioneered the use of coal in synthetic dyes and plastics.

The switch to oil began in the United States, which had abundant domestic oil and natural gas reserves.[4] By 1950, half of the total U.S. production of organic chemicals was based on natural gas and oil; by 1960, the proportion was 88 percent. The switch came later, but as rapidly, in Western Europe. In the United Kingdom in 1949, for instance, 9 percent of total organic chemical production was based on oil and natural gas, and the proportion rose to 63 percent by 1962. In Germany, the first petrochemical plant was set up in the mid-1950s, and by 1973, German companies derived 90 percent of their chemical feedstocks from oil. In light of Germany's substantial endowment of coal and the fact that German firms had made large, irretrievable investments in coal-related technologies, the changeover of the German industry is especially remarkable (Stokes 1988). The swift diffusion of petrochemical technology in Europe ironically was aided by the World War II destruction of the German and French industrial capacity, which reduced switching costs. However, the ready availability of petrochemical technology from American firms, especially by specialized suppliers of technology and engineering services, was crucial. Equally important, if not more so, was the postwar availability of crude oil from the Middle East, with the supplies

---

4. The U.S. and European statistics detailed in this passage are from Chapman (1991).

guaranteed by Pax Americana.[5] In other words, markets in both goods and technology played a fundamental role in facilitating the switch from coal to petroleum in the United States and Western Europe.

The enormity of this shift may not be easily appreciated today, until we begin to consider an analogous energy innovation challenge, such as a possible increase in the use of biomass. At that point, a range of issues analogous to the coal-to-oil switch arise. For instances, can processes be developed that can handle variations in the composition of biomass sources? Can processes be developed to remove impurities from biomass that might interfere with their processing (such as poisoning the catalyst)? Can the likely undesirable by-products be removed after processing in a cost-effective and scalable manner? Can existing equipment be used? Similar questions were confronted and solved through innovation and diffusion during the coal-to-oil shift.

This shift is also remarkable because direct government intervention was not central to the shift. For the most part, advances in oil refining, advances in how new types of chemical raw material could be extracted from crude oil and natural gas, and the diffusion of these advances did not rely upon government subsidies or other types of government incentives. Instead, these were in responses to changes in the availability of different types of resources and fundamental advances in the underlying scientific and engineering knowledge about materials and large-scale chemical reactions. Further, as we discuss at greater length later, the diffusion of these new technologies was largely mediated by the market for technology, in which specialized engineering firms played an important role.

This episode also illustrates the power of demand in calling forth the development of new technologies. As oil became more widely available, and as advances in oil-refining increased the potential for extracting more useful raw materials from crude oil and natural gas, there were incentives to explore oil as the source of the basic building blocks for chemical products. These incentives were especially marked in countries that were relatively abundant in oil. Western Europe was abundant in coal, but oil was relatively scarce. America was generously endowed with both coal and oil but, relative to Europe, was relatively abundant in oil and, moreover, had less invested in producing chemicals from inputs derived from coal. Not surprisingly, America led the shift from coal to oil.

---

5. The discovery of oil in the Middle East was significant not only because it meant that more oil was available, but also because the additional reserves would be supplied through a global market, and, therefore, control over oil was unlikely to be used as an economic or military weapon. The U.S. guarantee of unhindered oil supply to Western Europe was also very important in this respect. We are grateful to David Mowery for bringing this to our attention.

3.2.2   The Synthetic Rubber Research Program

The Synthetic Rubber Research Program offers an interesting instance of a government-funded cooperative research program in the chemical industry.[6] This wartime program took place from 1939 to 1945, when the United States faced being cut off from nearly 90 percent of the world's natural rubber supplies located in Southeast Asia. As a response, the U.S. government formed a consortium of the leading rubber firms, some of the leading chemical firms, and selected university researchers to expand and improve the production of synthetic rubber. The program was later extended to 1956. Between 1942 and 1956, the U.S. government invested $56 million for R&D in this consortium. The program mandated free exchange of information among participants.[7]

The principal objectives of the program were threefold. The first was to expand greatly the scale of synthetic rubber production. The second was to improve the quality of the synthetic rubber and produce specialty rubbers, such as rubbers suitable for use at low temperatures. The third objective, less explicit and of lower priority, was to develop greater knowledge and understanding of polymers.

The first objective of the program was successfully met. By 1945, the United States produced nearly 850,000 tons of synthetic rubber (up from less than 10,000 tons in 1941), more than seven times the peak German production achieved in 1943. Virtually all of the ramped-up production went to the war effort; the U.S. government purchased all production from the program, and companies producing synthetic rubber did not have to worry about being able to sell it. Having a guaranteed market in the U.S. government was important because it released the companies from concerns that there would be insufficient demand for their efforts.

The tremendous increase in the production of synthetic rubber did not require radical technological advance. Although a variety of alternative monomers (building blocks) were tried out, it turned out that butadiene and styrene, used in the Buna-S rubber patented in Germany in 1921, were the most suitable.[8] Neither was the basic process fundamentally new. Even so, a variety of logistical and technological problems had to be solved, such as expanding the supply of butadiene, for which advances in chemical engineering and petroleum refining were critical. Butadiene could be produced from the by-products of oil refining or natural gas. It could also be produced from other sources, such as industrial alcohol (ethanol). Many different sources of ethanol were considered, including alcohol from grain

6. This section draws upon Arora (2002). In addition, see Herbert and Bisio (1985) and Morris (1989) for detailed accounts and assessments of the program.
7. Eventually, thirty-six firms joined the consortium. In some cases, firms took special precautions to ensure that related internal projects were kept separate.
8. Buna-S rubber was called GR-S rubber in the United States.

or molasses. In Europe, Chemische Werke Huls (a German specialty chemicals company) nearly built a tire plant with French red wine as an ethylene source; the United States actually imported beet ethanol for butadiene for the synthetic rubber program shortly after World War II. During the tenure of the Synthetic Rubber Program, Congress passed laws mandating use of grain ethanol for butadiene production. *Plus ca change, plus c'est la meme chose!* The vast bulk of butadiene, however, was ultimately produced from oil-refining by-products.

Styrene, the other key input, could also be obtained from coal by-products, but oil-based sources were more promising, particularly after new processes were developed that could produce benzene ring type molecules. Government coordination played an important role in styrene production as well. The Dow Chemical Company was asked to take charge of styrene production for the program and build large styrene plants in Torrance (California), Velasco (Texas), and Sarnia (Ontario) using Dow technology and to supervise construction of several other plants. Union Carbide Corporation was asked to build a large plant based on its technology, and Monsanto Company was asked to build an ethylbenzene-styrene plant using Dow and Monsanto technology, in partnership with Lummus Corporation. The Koppers Company was asked to build a plant using vapor phase alkylation process using technical contributions from UOP, Phillips Petroleum Company, and Koppers. U.S. production rose from less than 2 million pounds per month in 1941 to 20 million pounds per month by the end of 1943. By the end of 1944, production was 40 million pounds per month (Spitz 1988).

The program's second objective of improving the quality of synthetic rubber was complementary to the first objective. Once again, though not fundamental in terms of the science involved, a number of improvements to the polymerization process were critical in producing synthetic rubber of a higher and more consistent quality. These changes were especially important for "cold rubber," a new type of synthetic rubber also made from butadiene and styrene but of higher quality in some ways. Among other uses, cold rubber was better suited for automobile tires. By 1954, two thirds of all synthetic rubber produced in the United States was "cold." Especially noteworthy improvements involved better control over gel formation, improved modifiers and emulsifiers, and, in the case of cold rubber, a new catalyst as well.

After World War II, the Synthetic Rubber Program was no longer the only or even the principal source of innovation in synthetic rubber. For example, oil-extended rubber—cold rubber with added mineral oils, which was both cheaper and easier to process than cold rubber—was developed by Goodyear Tire & Rubber Company and, independently, by General Tire, which was not part of the research program.[9] New modifiers and carbon black (an

---

9. There were benefits, nonetheless, from Goodyear's discovery, because Goodyear, as a member of the cooperative research program, made its discovery available to all other members.

important rubber additive) came from Phillips Petroleum, not a member of the program until after those discoveries. Similarly, the next generation of synthetic rubbers (the so-called nitrile rubber) were privately developed by two firms, Goodrich and Goodyear, outside the synthetic rubber program.

Although the government's research program succeeded in its two main objectives, the program had significant design flaws that ultimately narrowed its effectiveness in encouraging more widespread innovation. Perhaps its greatest flaw was its inflexible insistence on a common "recipe" to be followed by all participating companies; this may have delayed the adoption of a number of improvements (Morris 1989). The justification for the common-recipe requirement was that it would facilitate cooperation among members and simplify processing. However, according to Morris (1989), the effect was to complicate testing and large-scale introduction of new recipes. The common-recipe requirement also pushed many consortium members to spin off or partition some of their internal research projects, thus protecting themselves from the restrictions of the government program.

The information-sharing mandate was both a benefit and a curse for the program. On the one hand, the mandate allowed a number of firms to enter the industry, particularly after the end of World War II. This greatly increased competition and contributed to the rapid growth of the industry. On the other hand, the mandate diluted the incentives of firms to develop innovations because they would have to share the information with their rivals, lowering their private returns. Consequently, some participating firms had research groups working in rubber and polymer technologies that were carefully kept apart from the research groups participating in the Synthetic Rubber Program. Many of the major advances in technology came from outside the research sponsored by the program during that time period.

The greatest failure of the government research program in promoting fundamental technical advance (as opposed to increasing the production of synthetic rubber for the war effort), in Morris's opinion, was in the case of synthetic natural rubber (i.e., a synthetic route to the production of natural rubber, as opposed to a synthetic material with properties like those of natural rubber). Although GR-S (Government Rubber Styrene, formerly Buna-S) rubber worked well for automobile tire track, natural rubber was much better for truck and airplane tires, where loads and temperatures are markedly higher. Many program participants, from both industry and academia, tried to develop a suitable synthetic rubber for trucks and planes. The solution turned out to be a process of synthesizing natural rubber itself by polymerizing isoprene. The pioneering research of the German chemist Karl Ziegler in 1950 (not the U.S. government research program), in the catalytic polymerization of polyethylene and related materials, was instrumental in synthesizing natural rubber. Morris (1989) suggests that the information sharing requirements imposed by the government program may have delayed the application of Ziegler's discoveries until 1955.

Thus, neither the basic process for producing synthetic rubber, namely, the butadiene-styrene process, nor some of the important postwar improvements in the process for synthetic rubber can be credited to the government-funded program. The government research program succeeded in its immediate wartime aims, but not in its longer-term peace time ones. However, the government funds at stake were small. Between 1943 and 1955, only $55 million were invested in R&D, a fraction of the nearly $700 million capital investment in synthetic rubber production in 1945 alone and less than 2 percent of the total value of the synthetic rubber produced over this period.

The government's synthetic rubber research program succeeded in increasing production because it quickly settled on a viable technology, standardized the inputs (for the most part), and coordinated the production. As well, it coordinated with the users of synthetic rubber and provided capital to construct the various production facilities. The program worked best where the technology required incremental improvements advancements, which, nonetheless, cumulatively had a large impact. The program was able to coordinate private parties on a given standard (e.g., the GR-S rubber standard), which was desirable because the standard itself was a good one. Getting to a good standard was more likely because the technology was relatively mature and progress required was incremental.

Conversely, the Synthetic Rubber Program did not facilitate development of radical breakthroughs or fundamental discoveries in polymer science (although it contributed to both). This undoubtedly had much to do with the priorities of the program, which imposed certain inflexibilities that hindered radical innovation. In some cases, the information-sharing mandate diluted the incentives of firms to develop innovations because they would have to share the information with their rivals. Further, the strong applied orientation of the program implied a reduced priority for the training of graduate students, resulting in a much smaller contribution to the growth of polymer science than might otherwise have been.

The program's history also highlights the diversity of the potential sources of innovation. In the rubber research program, important innovations came from unexpected sources outside the program. General Tire invented "cold rubber," and Philips Petroleum, a smaller oil company from Oklahoma, albeit one with a strong tradition of chemical research, made significant contributions to the development of carbon black and other modifiers. Therefore, an innovation environment that has left ample room for new players has also made greater room for success, rather than one placing large early bets on a handful of players.

A third lesson from the history of the Synthetic Rubber Program is that of the role of demand, or procurement. During the war, all production from the program was purchased by the government at guaranteed prices, and companies were, therefore, not concerned about being able to sell their product. After the war, the renewed availability of natural rubber raised doubts about the viability of synthetic rubber, but the greater versatility of different types of synthetic rubber eventually prevailed.

3.2.3   The Synfuel Program

As early as the 1920s, the U.S. government had experimented with lique-faction of coal and other substances for fuels, but it did not result in a sig-nificant program until the end of World War II.[10] The 1944 Synthetic Liquid Fuels Act led to a $30 million five-year Department of the Interior research program that attempted to alleviate shortfalls in the supply of oil during after immediately after the war. The basic technologies, coal liquefaction and coal gasification, already existed, but neither was commercially viable at prevailing prices. The objective of the program was to promote new pro-cesses, catalysts, and input sources that would lead to commercial viability. Through the program, a demonstration coal hydrogenation plant was con-structed in 1949 that produced synthetic diesel and inspired great optimism for the prospect of synthetic fuel to replace fuel from crude oil, supplies of which at the time were limited and expensive. With a view of avoiding an "energy crisis" and dependence on "foreign oil" (a newly coined phrase at that point), Congress extended the program twice for an additional eleven years and increased funding to $87.6 million. However, the next plants built could not produce fuel at such low costs and high volume that had earlier excited so much optimism in the industry.

In the 1950s, a combination of lowered expectations for the synthetic fuels industry, the opening of Middle Eastern oil fields, new private forays into coal hydrogenation, and a political shift in both the executive and legislative branches of the U.S. government, led to Congress ending funding for syn-thetic fuel programs in 1953. Low-level research continued in "backburner" mode by the interior department, and later this research was transferred to the newly created Department of Energy in the 1970s.

The increases in crude oil prices in the 1970s revived the synthetic fuel efforts, with a mixture of motives consisting of reducing dependence on imported oil and promoting the use of coal, especially coal mined in the eastern part of the United States. The Energy Security Act of 1980 estab-lished the Synthetic Fuel Corporation (SFC), which was set up to promote the commercialization of these technologies. The corporation was a quasi-governmental entity comprising private industry partners and initially authorized for a maximum of twelve years and a maximum of $88 billion. The interest in synfuels heated up throughout the 1970s, and by the end of that decade, synfuels research was a major component of the nation's energy program. Spurred by instability in the Middle East, the high price of crude oil and fears that oil supplies were drying up, projected dramatic

---

10. This section relies upon Cohen and Noll's (1991) discussion of the program as well as http://fossil.energy.gov/aboutus/history/index.html and http://www.fas.org/sgp/crs/misc/RL33359.pdf.

increases in future oil costs, and concern that "market failures" were inhibiting private companies to invest in research, the synfuels program again regained popularity.

Under the program, some technical progress was made and large-scale processes were developed. Coal gasification appears to have been the most successful of the technological investments. One facility, the Cool Water Coal Gasification plant, which was provided price guarantees worth $115 million, succeeding in coming under budget and meeting the expected production and quality standards. However, Cool Water was a rare success for the SFC. Ultimately, low crude oil prices reduced the economic and strategic imperative to develop synthetic fuels from coal or other sources. The SFC was unable to lay the foundations for a commercially viable industry, in part because the technology failed to support reliable and large-scale production of affordable synthetic fuels. The broad consensus is that the government's synfuel programs failed, though there is less agreement on the causes of the failure.

At the risk of oversimplifying the contrast, the Synthetic Rubber Program succeeded because there was an enormous growth in demand for synthetic rubber, whereas the synthetic fuel program largely operated in an era of abundant, though occasionally expensive, oil and gas. Although price signals and signals about the future growth in demand are not the only ones that matter, they surely were very important for the failure of the synfuels program.

3.3   Diffusion and the Critical Role of Specialized Engineering Firms

In 1999, the world's then largest greenfield refinery and associated chemical plants, embodying the latest technology available with an estimated cost of around $6 billion, was commissioned in Jamnagar, India. The event passed by with little comment. In marked contrast, when an Indian company announced the development of a new car nearly a decade later, there was extensive publicity. And yet refining technology was developed only in the 1920s, whereas the internal combustion engine is nearly a century and a half old. One can infer that chemical technology has diffused so broadly that it is considered unremarkable for the latest technology to be used in a poor country. Put differently, there is no need for an Indian firm to try to design its own refineries and chemical plants because the required technology can be readily acquired at a reasonable price. The rapid diffusion of innovative techniques and processes is one of the striking features of the chemical industry. This is particularly true of chemical innovations since World War II, the diffusion of which were fostered by SEFs.[11]

---

11. This rapid diffusion greatly accelerated the maturation of the chemical industry and the consequent wide-ranging restructuring that the industry underwent in the United States in the 1980s, followed shortly by a similar restructuring in Europe (Arora and Gambardella 1998).

In a pioneering study, Freeman (1968, 30) noted that for the period 1960 to 1966, "nearly three quarters of the major new plants were 'engineered,' procured, and constructed by specialist plant contractors." Moreover, Freeman found that SEFs were an important source of process technologies; during 1960 to 1966, they accounted as a group for about 30 percent of all licenses of chemical processes. Freeman's findings are confirmed by more recent data analyzed in Arora and Gambardella (1998). These data show that for the period 1980 to 1990, almost three-fourths of all chemical plants in the world were engineered by SEFs. Although the share of SEFs varies across different types of chemical products, in practically all of them it is above 50 percent. Moreover, SEFs still account for about 35 percent of all licenses. However, SEFs in the 1980s were perhaps more important than this figure suggests because their very presence induced many downstream companies to license their processes. In essence, SEFs helped create a market for technology, making process technology into a commodity that could be bought and sold.

The first SEFs were formed in the early part of the twentieth century, and their clients were typically oil companies. However, SEFs also started operating in bulk chemicals such as sulfuric acid and ammonia. Later, most SEFs would design large-scale plants for refineries and petrochemical building blocks, such as ethylene. Companies such as Kellogg Engineering, Badger Engineering, Stone and Webster, and UOP are prominent examples of early SEFs.

In addition to diffusion, the SEFs also played a crucial role in the development of new and improved processes. However, with a few notable exceptions, SEFs did not develop radically new processes. The contributions of SEFs have been largely in two areas: catalytic processes and engineering design improvements.[12] For the most part, however, major process innovations came from the large oil and chemical companies (Mansfield et al. 1977). The SEFs were most effective at moving new processes down the learning curve. By acting as independent licensers, SEFs also induced chemical firms to license their technology (Arora and Gambardella 1998). The SEFs provide a vivid illustration of the economies of scale operating at the industry level, rather than at the plant or individual firm. By specializing in process design and engineering and by working for a number of clients, SEFs could learn and accumulate skills and expertise that no single chemical company could match.[13]

Adam Smith noted more than 200 years ago that specialization and divi-

12. A number of other SEFs have contributed to advances in engineering design. For instance, Kellogg made significant contributions to developing high-pressure processes for ammonia in the 1930s, while Badger is associated with fluidized-bed catalytic processes (in collaboration with Sohio). Similarly, the Danish SEF, Halder Topsoe, is the leading source of technology for ammonia plants. Lurgi, a German firm, licenses a number of technologies, including coal gasification.

13. As independent developers of technology, SEFs are similar in some respects to today's biotech companies, allying with a number of chemical firms in developing new technologies.

sion of labor are more extensive in larger markets. As Bresnahan and Gambardella (1998) demonstrate, a large market is important for technology specialists, but this growth in market must be in the form of an increase in the breadth (more buyers) rather than merely depth (each buyer becomes larger). Indeed, SEFs account for a greater share of licenses to small firms and firms in developing countries, indicating the mutual dependence between specialist technology suppliers and firms that are not technically sophisticated (figure 3.2).

The challenge of innovation is more than the mere discovery and development of new technologies and feedstocks. The technology and processes for using these alternatives will have an impact on the environment only if they are broadly diffused and used. Technology diffusion has been faster when firms like SEFs have played a role in offering technology, sometimes bundled with engineering services. Rapid diffusion of innovation in the chemical industry has resulted when major innovations were not concentrated in the hands of a few firms, no matter how innovative. Instead, multiple sources of innovations appear to have been desirable.

3.4   Innovation Policies in the Chemical Industry

Two policy areas stand out with respect to their effect on innovation in the chemical sector. First, intellectual property rights protection, particularly patents, has stimulated innovation and the diffusion of innovations. Second, licensing and antitrust regulations have fostered wider competition and sped up diffusion of technologies.

3.4.1   Intellectual Property Rights

The chemical industry has seen extensive technology diffusion though it also enjoys strong patent protection. Although this may sound like a contradiction in terms, it is not. In chemicals, although patents are effective in deterring straight imitation, rivals were often able to develop competing variants of patented chemical processes. These processes differed in terms of operating conditions, starting materials, or in terms of yields, conversion rates, and properties of the final material. The ability to develop different processes resulted in a vigorous competition in the market for technology and a rapid diffusion of new technologies.

Despite the importance of technology specialists, the chemical industry has largely been spared the problems that some nonmanufacturing patent holders have created in several information technology sectors. Patents largely work well in the chemical industry in that they encourage the invention of new technologies and follow-on investment in commercializing those technologies (Cohen, Nelson, and Walsh 2000). They also encourage the widespread use of new technologies. Chemical patents work better because they are less "fuzzy." They are less fuzzy because the object of discovery can be described clearly in terms of formulas, reaction pathways, operating conditions, and the like (Levin et al. 1987). But it is not merely that the object of discovery is more discrete in the sense of being a particular compound. Rather, it is the ability to relate the "essential" structure of the compound to its function. This allows a patent to include within its ambit inessential variations in structure, as in minor modifications in side chains of a pesticide.[14] In fact, chemical product patents frequently use Markush structures, which permit a succinct and compact description of the claims and allow the inventor to protect the invention for sets of related compounds without the expense (and tedium) of testing and listing the entire set.[15] The ability to explicate the underlying scientific basis of the innovation allows the scope of the patent to be delimited more clearly. The obvious extensions can be foreseen more easily and described more compactly.

Conversely, when innovations cannot be described in terms of universal and general categories, sensible patent law can only provide narrow patent protection. Failure to do so results in costly patent disputes, sometimes with devastating consequences for the economy, as an early epoch in the history of the chemical industry itself shows. During the 1860s, for example, when synthetic dyes first appeared, their structure was poorly understood. Broad patents led to litigation and, in some cases, unwarranted and harmful monopolies. In France, an excessively broad patent on *Fuchsine* (aniline red) was construed to include all processes for making the red aniline-based dye, even though the structure of aniline dyes was as yet unknown. There were also long and bitter disputes in England about the validity of the Medlock patent for *magenta* (another aniline dye); the dispute turned on the appropriate definition of "dry" arsenic acid (i.e., with or without water of crystallization). In the case of *aniline blue,* the dispute rested on whether the substitution of an organic acid for an inorganic acid in the production process was enough to avoid infringement (Travis 1993). The British courts interpreted patents narrowly, with the result that competition in the British organic dyestuffs industry remained vigorous until the industry itself was overwhelmed by its German rivals. By contrast, the *Fuchsine* monopoly in France devastated the local industry. It should be recalled that France at first rivaled Germany in organic chemistry, but the French synthetic dye industry greatly suffered from the *Fuchsine* dispute.

Until World War II, chemical producers did use patents to restrict entry and carve up markets. However, after the war, patents were used more frequently to facilitate technology licensing, particularly outside home mar-

14. In some instances, seemingly minor variations in side chains can have significant biological effects. Therefore, what is a "minor" variation is itself determined by the state of the current understanding of the relation between structure and function.

15. A Markush structure is best understood as a language for specifying chemical structures of compounds, which allows generic representation for an entire set of related compounds. See Maynard and Peters (1991) for details.

kets. In a marked departure from their pre–World War II strategy of closely controlling their technology, a number of chemical and oil companies began to use licensing as an important (although not the only) means of profiting from innovation. Spitz (1988, 318) describes the licensing of the Hercules-Distillers phenol/acetone process "to any and all comers"; the process "was commercialized in 1953 and forever changed the way that phenol would be produced." This remarkable change in behavior appears to have been triggered in part by a newly vigorous antitrust policy in the United States, an issue that we discuss in some depth in the following.

One reason why patent protection promoted diffusion of technology was the greater competition after World War II in both product markets and especially in the market for technology. When multiple sources of technology exist, then even when the technology sources are not exact substitutes, they can provide effective competition. Arora and Fosfuri (2003) formally show that the presence of other technology holders, especially those that do not participate in the product market, encourages licensing, facilitating entry downstream.

It is also noteworthy that SEF licensing activities appear not to have been hindered by patenting. Indeed, in processes in which high rates of patenting occur, SEFs are more active than in processes with low rates of patenting (figure 3.1). Insofar as patenting rates are indicators of technological activity, this suggests that SEFs are active in diffusing technologies not simply in mature sectors, but also in sectors with high rates of technical advance.

An active technology market has also encouraged new entrants into the product market. Arora, Fosfuri, and Gambardella (2001) show that markets for technology encourage investment by chemical firms in developing countries, implying that technology suppliers lower entry barriers. Lieberman (1989) finds that licensing was less common in concentrated chemical products, and the limited licensing that did take place was by outsiders (nonproducers and foreign firms). Moreover, he finds that when licensing was restricted, there was less entry. In a related study of twenty-four chemical product markets, Lieberman (1987) reports that patenting by outsiders was associated with a faster decline of product price, once again suggesting that outside patenting encouraged entry in the product market. This is borne out by Arora and Fosfuri (2000), who show that the principle source of demand for technology licensed by SEFs are small firms in Western markets (North America, Japan, Western Europe) and firms in developing countries (figure 3.2).

3.4.2   Licensing and Antitrust Regulations

As we noted in the preceding, post–World War II, chemical firms have tended to license their technology, which has greatly contributed to technology diffusion. Ralph Landau observed in 1966 that the "the partial breakdown of secrecy barriers in the chemical industry is increasing . . .

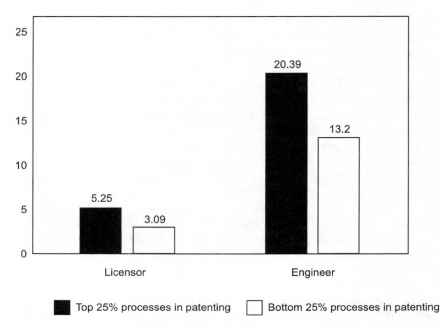

Fig. 3.1   Average number of specialized engineering firms by patent intensity in process, 139 process technologies (1980–1990)

Source: Our calculations, based on Chemical Age Project File (1991) data.

Note: The top 25 percent processes are in the top quartile in terms of the number of patents, and the bottom 25 percent of the processes are in the last quartile in terms of the number of patents filed.

the trend toward more licensing of processes" (Landau 1966, 4). Chemical firms have licensed heavily because they have faced competition in both the product and the technology markets. These two types of competition are interrelated. A competitive product market will encourage the entry of technology specialists (Bresnahan and Gambardella 1998). When Standard Oil, with its dominant position in the oil market, tried to restrict access to refining technology, the independent oil firms turned to specialized technology suppliers such as UOP. More generally, at crucial stages in the industry's history, antitrust rulings have directly increased competition in the product market and also reduced concentration of technology ownership, increasing competition in the market for technology.

The two firms prominently featured in the context of antitrust enforcement in the United States were Standard Oil and DuPont. William Burton, a scientist at Standard Oil developed the first commercially successful cracking process, a first major innovation in refining technology, in 1909 to 1910. However, Standard Oil was reluctant to invest in the process. As a result of an antitrust suit, the original Standard Oil was broken up into several firms

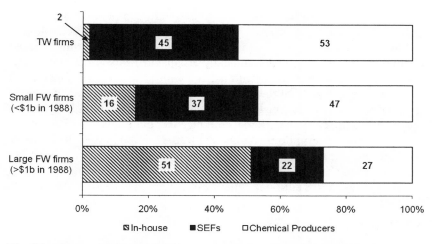

Fig. 3.2   Share of specialized engineering firms in technology licensing by type of buyer

Source: Arora and Fosfuri (2000).

Notes: TW = third world. Includes all countries except those in Western Europe, Japan, North America, Australia, and Eastern Europe (except Soviet Bloc). FW = first world. Comprises Western Europe, Japan, North America, and Australia. Large FW = first-world firms with turnover greater than $1 billion in 1988. Small FW = all other first-world firms.

in 1911, among which was Standard Oil of Indiana, where Burton worked, and which not only commercialized the process but also licensed it to a number of other oil refiners. The high royalty charged by Standard of Indiana led others to develop alternative processes, among which was UOP, which eventually developed into a leading supplier of technology to the petroleum refining industry.

The case of DuPont provides another important example of the role of antitrust policy. Founded as a maker of explosives powder, DuPont was split into three separate firms following a successful antitrust suit in 1913. The antitrust suit convinced the management of DuPont that the only path to future growth lay in entering new markets through innovation rather than acquisition of existing producers. In 1926, DuPont signed a comprehensive technology sharing agreement with Imperial Chemical Industries (ICI), which also involved market-sharing.[16] The agreement entitled DuPont to exclusive access in the United States to ICI technology, such as polyethylene. However, the fear of antitrust action pushed ICI to license polyethylene to other firms as well after World War II. Similarly, DuPont was nudged into licensing its nylon technology (and know-how) to a Monsanto joint venture, Chemstrand, in 1951. More recently, antitrust authorities in the United States and Europe intervened when Dow Chemical acquired Union Carbide.

16. The agreement would be dissolved in 1952.

The antitrust ruling attempted to try to maintain competition both in the polyethylene market as well as in the market for polyethylene technology.[17]

The more general point is that licensing flourished when firms faced competition from other technologically capable firms (whether at home or abroad) and licensing itself facilitated entry into the industry. The second tentative conclusion from the history of the chemical industry is that anti-trust enforcement, including the occasional episodes of compulsory licensing, does not appear to have had a chilling effect on innovation. Part of the reason that innovation flourished is that in the United States, the industry has had multiple sources of innovation. No single firm, not even DuPont, has dominated the chemical industry, in the way I.G. Farben and ICI dominated their respective national industries between the two world wars.

3.5   Conclusions and Caveats

Three noteworthy aspects of innovation characterize the chemical industry.[18] First, chemical innovations are deeply rooted in science. However, despite a worldwide tradition of government support for scientific research, chemical R&D has been largely privately funded. In the United States, federal government support for chemical research, already at a relatively low level, has steadily declined over time. Second, other than early in the history of the industry when new dyes relied upon the scientific advances in organic chemistry led by German chemists, innovations have largely come from firms rather than universities. Universities, on the other hand, have played an important role in institutionalizing the learning, creating new disciplines (which have been crucial for sustaining innovations), and developing human capital. Third, the major phases of chemical innovation have been accompanied by large growth in demand.

These three aspects are closely related. The chemical industry rose to prominence when government support for research was uncommon, chemical innovations could capture large markets by substituting for a variety of materials used as industrial inputs, and the scientific understanding linking the structure of their materials to their properties could increase the productivity of attempts to discover new and useful materials. Thus, it made commercial sense for firms to invest not simply in applied research, but also in basic research—an area typically the domain of universities and government programs. Therefore, chemical innovation has relied more heavily than some other innovation-based industries upon private investments in research, development, and commercialization. Correspondingly,

---

17. Dow Chemical and British Petroleum (BP) Amoco were competing against Univation Technologies, a joint venture of ExxonMobil and Union Carbide. Dow and Exxon held the basic catalyst patents, and BP and Union Carbide supplied process technology for polyethylene. As a condition of the acquisition, Dow made its catalyst technology available to BP-Amoco.

18. The beginning of this section draws upon Arora (2002).

government supported research, and university discoveries, have been less important.

However, although they feature centrally, established firms have not been the only actors in chemical innovations. Rather, they share the stage with a variety of other firms, including start-ups. If nylon and polyethylene were discovered in corporate labs (of DuPont and ICI, respectively), polyester, the most successful synthetic material used in everything from clothes to plastic bottles, was discovered by chemists working for the Calico Printers Association, a group mostly concerned with textile printing. The fundamental advances in catalysts for producing polyethylene and polypropylene came from the work of Karl Ziegler, a German chemist funded by the local coal industry association, and from a small oil company in Oklahoma, Phillips. If the first major refining technology, thermal cracking, originated from Standard Oil, the next one, catalytic cracking, was invented in 1936 by Eugene Houdry, a French engineer who moved to the United States to commercialize his invention. The technology was significantly improved by a group of firms led by Standard Oil of New Jersey. The implication is that innovation in the chemical industry has drawn upon a diverse range of sources, including corporate R&D labs and government programs, but also a variety of small firms and start-ups.

Large government initiatives have had mixed success. They have been successful in coordinating private decisions when innovation required complementary improvements in inputs and uses, as well as large investments in complementary infrastructure. They have been successful in coaxing incremental technical advances, which have cumulatively contributed significantly to productivity growth. The record of such initiatives in producing significant new technical advances in the chemical industry is poor, in part because these initiatives have had conflicting goals, such as increasing the efficiency of existing technology along with the development of new technologies.

The history of the chemical industry shows that technology has diffused effectively through markets for technology, without need for direct government subsidies. Markets for technology have also offered a prominent role for start-ups and other types of technology specialists, such as SEFs, which have been the engine of diffusion for chemical innovation as well as a frequent contributor. The chemical industry's history shows that indirect government policies that promote competition in the product market as well as in the market for technology can promote technical advance and productivity growth.

Competition in the technology market does not imply weakening intellectual property protection; rather, strong patent protection can facilitate competition by encouraging innovation by firms outside the industry, including start-ups. However, broad patent protection is effective when the underlying knowledge base is strong. For new bodies of knowledge, narrowly crafted patent protection works better at encouraging innovation and preventing logjams. In addition, from time to time, antitrust policy has prevented chemical technology ownership from being concentrated in a few hands, enhancing competition among technology holders.

History rarely repeats itself, and its lessons cannot be applied mechanically. However, it appears that as was the case of chemical innovation, energy innovation is more likely to be successful and effective when private R&D from diverse sources is stimulated and strong patents protection is combined with robust antitrust enforcement.

References

Arora, Ashish. 2002. The government's role in chemical technology innovation, then and now. In *The role of government in energy technology innovation: Insights for government policy in the energy sector,* ed. Vicki Norberg-Bohm, 109–26. Belfer Center for Science and International Affairs, John F. Kennedy School of Government, Harvard University, Working Paper.

Arora, Ashish, and Andrea Fosfuri. 2000. The market for technology in the chemical industry: Causes and consequences. *Revue d'Economie Industrielle* 92:317–34.

———. 2003. Licensing the market for technology. *Journal of Economic Behavior and Organization* 52 (2): 277–95.

Arora, Ashish, Andrea Fosfuri, and Alfonso Gambardella. 2001. Specialized technology suppliers, international spillovers and investment: Evidence from the chemical industry. *Journal of Development Economics* 65 (1): 31–54.

Arora, Ashish, and Alfonso Gambardella. 1998. Evolution of industry structure in the chemical industry. In *Chemicals and long-term economic growth,* ed. Ashish Arora, Ralph Landau, and Nathan Rosenberg, 379–414. New York: Wiley.

Arora, Ashish, and Nathan Rosenberg. 1998. Chemicals: A U.S. success story. In *Chemicals and long-term economic growth,* ed. Ashish Arora, Ralph Landau, and Nathan Rosenberg, 71–102. New York: Wiley.

Bresnahan, Timothy, and Alfonso Gambardella. 1998. The division of inventive labor and the extent of the market. In *General-purpose technologies and economic growth,* ed. Elhanan Helpman, 253–67. Cambridge, MA: MIT Press.

Bresnahan, Timothy, and Shane Greenstein. 1996. Technical progress and co-invention in computing and in the uses of computers. *Brookings Papers on Economic Activity, Microeconomics:* 1–83.

Chapman, Keith. 1991. *The international petrochemical industry: Evolution and location.* Oxford, UK: Basil Blackwell.

Chemical Age Project File. 1991. Database available from Reed Telepublishing. London: Reed Elsevier. http://library.dialog.com/bluesheets/html/b10318.html.

Cohen, Linda, and Roger Noll. 1991. Synthetic fuels from coal. In *The technology pork barrel,* ed. Linda Cohen and Roger Noll, 259–320. Washington, DC: Brookings Institution.

Cohen, Wesley M., Richard R. Nelson, and John P. Walsh. 2000. Protecting their intellectual assets: Appropriability conditions and why U.S. manufacturing firms patent (or not). NBER Working Paper no. 7552. Cambridge, MA: National Bureau of Economic Research.

David, Paul A. 1990. The dynamo and the computer: An historical perspective on the modern productivity paradox. *American Economic Review* 80 (2): 355–61.

Freeman, Christofer. 1968. Chemical process plant: Innovation and the world market. *National Institute Economic Review* 45 (1): 29–51.

Herbert, Vernon, and Attilio Bisio. 1985. *Synthetic rubber: A project that had to succeed.* Westport, CT: Greenwood Press.

Hounshell, David A. 1988. The making of the synthetic fiber industry in the United States. Harvard Business School. Unpublished Manuscript.

Hounshell, David A., and John Kenly Smith, Jr. 1988. *Science and corporate strategy: DuPont R&D 1902–1980.* Cambridge, UK: Cambridge University Press.

Landau, Ralph. 1966. *The chemical plant: From process selection to commercial operation.* New York: Reinhold.

Landau, Ralph, and Nathan Rosenberg. 1992. Successful commercialization in the chemical process industries. In *Technology and the wealth of nations,* ed. Nathan Rosenberg, Ralph Landau, and David Mowery, 73–120. Stanford, CA: Stanford University Press.

Landes, David S. 1969. *The unbound Prometheus: Technological change and industrial development in Western Europe from 1750 to the present.* Cambridge, UK: Cambridge University Press.

Levin, Richard C., Alvin K. Klevorick, Richard R. Nelson, and Sidney G. Winter. 1987. Appropriating the returns from industrial research and development. *Brookings Papers on Economic Activity, Special Issue on Microeconomics:* 783–831.

Lieberman, Marvin B. 1987. Patents, learning by doing, and market structure in the chemical processing industries. *International Journal of Industrial Organization* 5 (3): 257–76.

———. 1989. The learning curve, technological barriers to entry, and competitive survival in the chemical processing industries. *Strategic Management Journal* 10 (5): 431–47.

Mansfield, Edwin, John Rapoport, Anthony Romeo, Edmond Villani, Samuel Wagner, and Frank Husic. 1977. *The production and application of new industrial technology.* New York: Norton.

Maynard, John T., and Howard M. Peters. 1991. *Understanding chemical patents: A guide for the inventor.* Washington, DC: American Chemical Society.

Morris, Peter J. T. 1989. *The American Synthetic Rubber Research Program.* Philadelphia, PA: University of Pennsylvania Press.

Murmann, Johann Peter. 2003. *Knowledge and competitive advantage: The coevolution of firms, technology, and national institutions.* Cambridge, UK: Cambridge University Press.

Rosenberg, Nathan. 1998. Technological change in chemicals: The role of university-industry relations. In *Chemicals and long-term economic growth,* ed. Ashish Arora, Ralph Landau, and Nathan Rosenberg. New York: Wiley.

Spitz, Peter H. 1988. *Petrochemicals: The rise of an industry.* New York: Wiley.

Stokes, Raymond G. 1988. *Divide and prosper: The heirs of I. G. Farben under Allied authority 1945–51.* Berkeley, CA: University of California Press.

Struben, Jeroen, and John Sterman. 2008. Transition challenges for alternative fuel vehicle and transportation systems. *Environment and Planning B: Planning and Design* 35 (6): 1070–97.

Travis, Anthony S. 1993. *The rainbow makers: The origins of the synthetic dyestuffs industry in Western Europe.* London: Associated University Presses.

Finding the Endless Frontier
Lessons from the Life Sciences Innovation System for Energy R&D

Iain M. Cockburn, Scott Stern, and Jack Zausner

4.1 Introduction

Over the past thirty years, the evolution and performance of the biopharmaceutical sector has been remarkable. Combined with related advances in medical practice, new pharmaceuticals and biologics have delivered extraordinary economic and health benefits, from drugs that offer proactive treatments against conditions such as hypertension to the development of life-saving drug regimens to combat acquired immune deficiency syndrome (AIDS) and cancer. Biopharmaceutical innovation has been linked to significant improvements in mortality and clinical outcomes (Lichtenberg 1998, 2008) and offers cost savings and productivity gains over alternative treatments (Garthwaite 2009). Moreover, the biopharmaceutical industry has been a reliable source of high-wage, high-skilled employment, and, until the recent past, has generated persistently high returns to investors (Burrill & Company 2008). Widely regarded as a success story for government support of research and development (R&D), this sector has attracted increasing policy attention as a potential source of regional and national economic development (Cortright and Mayer 2002; Feldman 2003; Hermans et al. 2008).

This chapter outlines some key features of the life sciences innovation

Iain M. Cockburn is professor of finance and economics at Boston University, and a research associate of the National Bureau of Economic Research. Scott Stern is professor of technological innovation, entrepreneurship, and strategic management at the Sloan School of Management, Massachusetts Institute Technology; and a research associate and director of the working group on Innovation Policy and the Economy at the National Bureau of Economic Research. Jack Zausner is an associate at McKinsey & Company.

We thank Adam Jaffe, Rebecca Henderson, and other participants in the preconference workshop for helpful comments. Megha Kahani provided valuable research assistance.

system—the set of interdependent firms, markets, institutions, and regulatory and legal frameworks responsible for this strong record of innovation—and draws some lessons from this sector for innovation policy in energy and climate change. While we recognize that the nature of energy and climate change innovation is in many respects different in some fundamental ways from life sciences innovation (and we discuss these differences in the following), we, nonetheless, argue that the evolution and structure of the life sciences innovation system offers an instructive comparison. Most notably, the genesis and evolution of the life sciences innovation system is the consequence of a set of policy choices and a microeconomic environment that has allowed the United States to leverage a set of embryonic scientific discoveries into a platform for sustained innovation, which has had a significant impact on human health and welfare.

By any measure, technological progress in biopharmaceuticals over the past thirty to forty years has been impressive. Figure 4.1 shows the decline in mortality in the United States in recent decades. While factors such as public health, nutrition, access to medical care, and progress in other medical tech-

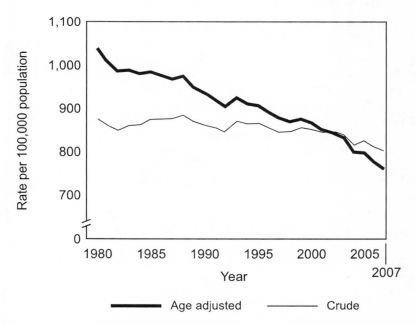

Fig. 4.1   Crude and age-adjusted death rates: United States, 1980–2006 final and 2007 preliminary

Sources: Drawn from Xu, Kochanek, and Tejada-Vera (2009). Center for Disease Control and Prevention/National Center for Health Statistics (CDC/NCHS), National Vital Statistics System, Mortality.

Note: Crude death rates on an annual basis are per 100,000 population; age-adjusted rates are per 100,000 U.S. standard population.

nologies have played an important role, much of this can be attributed to innovation in biopharmaceuticals, with major breakthroughs in treatment of leading causes of death such as heart disease, cancer, and human immunodeficiency virus (HIV).[1] Indeed, much innovation in this sector can be characterized as radical, with remarkable advances made in treating disease through the identification and exploitation of new physiological mechanisms or new classes of drugs, from selective serotin reuptake inhibitors (SSRIs) in the 1980s to HIV/AIDS drugs in the 1990s and genomic therapies in the current decade. But incremental innovation in the form of development of "follow-on" drugs with differentiated properties, or efforts to enhance the effectiveness of existing drugs through reformulations or more-effective treatment regimens, has also been pervasive and has been a significant source of economic and health benefits (Cockburn 2007). Though a low rate of U.S. Food and Drug Administration (FDA) approvals over the last several years has led to significant concern over the "biopharmaceutical R&D productivity crisis," most scientists and analysts forecast continued evolution and significant growth into the future: an unprecedented numbers of drug candidates are in the early stages of the clinical development process, and new cohorts of clinically relevant (and commercially profitable) new drugs and therapies are emerging from the significant basic research investments made over the past two decades in areas such as genomics and stem cells. Importantly, despite dramatic shifts in the nature of the underlying scientific base of the industry, changes in the nature and locus of demand (from cash payers to managed care), and realignments in both vertical and horizontal industry structure, the life sciences innovation system has, at the macro level, continued to grow and evolve over time.

While the commercialization of new drugs and therapies is by and large undertaken by the private sector, a distinctive attribute of the life sciences innovation system is that the biopharmaceutical industry draws upon (and complements) an exceptionally large publicly funded basic research effort in the life sciences. Life sciences now account for more than 60 percent of all academic R&D expenditures (National Science Foundation [NSF] 2008), with the vast majority of support for academic R&D coming from the budgets of the National Institutes of Health (NIH), NSF, and other agencies. These expenditures contribute to the development of a skilled and specialized life sciences workforce, a high rate of sustained advance in "upstream" science performed in academic and government laboratories, and the development of platform technologies, data sets, and research materials that serve as a foundation for commercial applications. In some areas such as cancer

1. While the link between innovation and improved disease outcomes is subtle (demographic and behavioral shifts play a very important role), there is considerable evidence that improvements in outcomes are closely linked to disease categories and conditions that have seen significant biomedical innovation. See Lichtenberg (1995, 2001, 2005) and Duggan and Evans (2008).

and viral outbreaks, the government also has been directly involved in the identification of new potential therapies and drugs.

Federal involvement in biopharmaceutical innovation has not been confined to underwriting the development of a body of scientific knowledge and an army of specialized human capital. The industry is highly regulated (arguably one of the most regulated sectors of the economy), with the FDA controlling new product introductions and production processes and federal and state legislation and regulation governing the marketing, distribution, and reimbursement rules for drugs. Significantly, while these regulatory structures are a significant, and costly, constraint on commercial innovation, they also play an important role in shaping competition. Though the high costs of meeting FDA requirements and managing the FDA approval process lowers the direct returns realized by any one successful drug, these costs (and strong patent protection) create significant barriers to entry; as a result, innovators are insulated from direct price competition, and competition can be premised on innovation and the improvement of patient care and product quality.

What are the forces driving innovation in this sector? How has the life sciences innovation system been able to yield such a high and sustained rate of scientific discovery and technological innovation? While scientific and technological opportunity in this sector has been very high in the aftermath of discoveries such as Watson and Crick's elucidation of the structure of DNA, the principal contention of the chapter is that the economic and technological dynamism of the life sciences sector over the past thirty years does not simply reflect unusual scientific opportunity. Rather, the performance of the life sciences innovation system is grounded in the microeconomic and institutional environment, which has, by and large, been conducive to long-run scientific and technical progress.

Our analysis focuses on six key elements. First, and perhaps most important, the life sciences innovation system is built on a stable foundation of high levels of public support for basic research over the long term. Public funding for academic biomedical research through the NIH, together with other granting agencies, has been sustained at a high level for many decades. While political considerations (and direct policy choices) have led to occasional "surges" in the overall NIH budget (as occurred from 1998 to 2003) and occasional rapid shifts in financial resources across research programs (such as during the focus on HIV in the late 1980s and early 1990s), as shown in figure 4.2, the NIH has, by and large, been able to maintain a relatively steady funding growth rate for both intra- and extramural research over long periods of time.

Second, innovation in the life sciences innovation system is grounded in the development of a large and specialized R&D workforce. Over time, universities have expanded and developed graduate and postgraduate training programs, particularly within academic medical centers, allowing each new

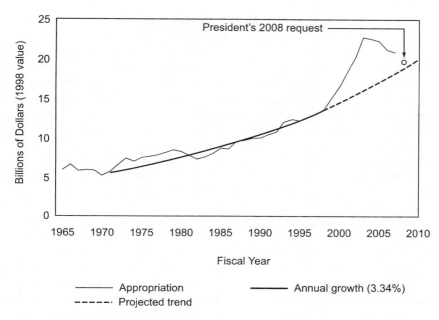

Fig. 4.2   Federal funding of NIH over time

generation of researchers to develop the specialized human capital required to innovate at the life sciences frontier. Federal and state policies have reinforced this emphasis on human capital investment through programs that specifically fund training and provide universities with incentives to conduct research that utilizes graduate and postdoctoral researchers. Over time, there has been a significant increase in both the public and private life sciences innovation workforce over time (NSF 2008). Moreover, this increase in size has been accompanied by a significant increase in the degree of specialization by individuals: while researchers share undergraduate training in a few core disciplines such as biology and chemistry, doctoral and postdoctoral training emphasizes the development of specialized expertise. Over time, the organization of both public and private research has evolved into (often quite large) collaborative research teams composed of highly specialized individual team members; together, these teams bring together knowledge, expertise, and specialized tools and materials from related (but distinct) scientific research fields.

Third, the nature of demand for biopharmaceuticals has had a profound impact on the life sciences innovation system. Intrinsically high willingness to pay for products that extend life or improve the quality of life has been translated into relatively price-inelastic and stable demand, increasingly controlled by insurers and government, but, nonetheless, driven to a large extent by physician and patient preferences. "Blockbuster" products have by and large been those which are the first to offer safe and effective treatment for poorly served medical conditions or those with demonstrably superior clinical attributes, rather than price-competitive alternatives to existing products. At the same time, government regulation of approval of generic products and of retail dispensing of drugs has supported a very efficient distribution channel for generics, creating a powerful "push" motive for R&D-based companies to innovate. Notwithstanding increasing emphasis by some companies on incremental improvement of existing products, or on marketing to expand their use, and concerns about whether payers will continue to reward innovative product introductions, biotechnology and pharmaceutical firms have been able to secure significant returns over a long period of time by focusing on the development and commercialization of innovative novel biotherapeutic compounds.

Fourth, notwithstanding the political priorities reflected in appropriations bills and occasional large-scale top-down initiatives, the allocation of public funding for biomedical research has been primarily focused on investigator-initiated, peer-reviewed projects. Most public research funds support extramural research, primarily conducted at universities, where researchers face explicit incentives and social norms that reward individual creativity and academic freedom. Alternative sources of support such as the Howard Hughes Medical Investigator program further reinforce an orientation in which the direction of research is controlled largely by researchers themselves, and the quality of research is judged through peer review. With relatively few exceptions, government-sponsored research has rarely taken the form of "Manhattan Project" initiatives. Instead, biomedical science has been largely driven by a robust and independent scientific community focused on the intellectual merit and novelty of investigator-generated research proposals, balanced by input concerning social or governmental priorities. Indeed, the most notable "big science" projects within the life sciences innovation system such as the Human Genome Project have been initiated within the scientific community (with the objective of providing a platform for investigator-initiated follow-on research projects).

Fifth, intellectual property rights (IPRs)—most notably in the form of patents—play a fundamental (if occasionally controversial) role in this sector, and there is considerable evidence that the patent system is relatively more effective in the life sciences than in many other areas of the economy. Relative to the formal IPRs available for other sectors, IPRs in biopharmaceuticals is relatively unambiguous, visible, and enforceable, and, by and large, is closely synchronized to product lifetimes and to relevant product market regulation. Biopharmaceutical firms rely heavily on strong IPRs to protect innovations during the product development process and during a period of time after FDA approval; in most circumstances, these innovators then face intense generic competition once the patent underlying a compound or treatment expires. The availability of IPRs for final products during the early years of the product life cycle and intense competition driving price to marginal cost after patent expiration simultaneously induces significant consumer welfare and powerful incentives for firms to launch new products over time. Over the last decade, a vigorous debate has emerged over the role of patents for more upstream discoveries in the life sciences. While this debate is ongoing, an emerging body of evidence suggests that the role of patents is quite complex; while it is possible to identify particular cases where patent grants have been associated with what seems to be significant inefficiencies, the more general pattern seems to be that, over time, strong patents operate in parallel (and are complementary to) a large and vigorous domain for open science. Notably, the patent system plays an important role in facilitating a market for technology that promotes the commercialization of publicly funded research discoveries.

Finally, the life sciences innovation system is characterized by intense competition on the basis of innovation. While price competition in the product market for biopharmaceuticals is relatively muted (at least until generic entry occurs after patent expiration), competition between researchers, institutions, and firms is focused on discovery, innovation, and the commercialization of new technologies. Individual scientific research teams compete with each other for scientific "kudos"; universities compete with each other to attract faculty, students, and resources; biotechnology firms compete with each other to attract scientists, venture capital, and commercialization partners; and product market competition is, by and large, oriented around quality and innovation rather than cost. In other words, despite FDA regulation and the presence of strong patents, competition within the life sciences innovation system is pervasive and operates at multiple levels and at different stages of the product development process.

In concert, these drivers seem to have been instrumental in shaping the structure and evolution of the life sciences innovation system. To draw out the lessons (and points of difference) for energy and climate change innovation, we begin in the next two sections with an historical overview. This historical narrative emphasizes that the evolution of the life sciences innovation system over time does not simply track shifts in scientific or technological opportunity, but instead reflects specific episodes and instances of institutional and economic experimentation. We use this background to then analyze in greater detail the six distinctive characteristics of the now-mature life sciences innovation system previewed in the preceding. The final section of the paper draws out the lessons from the life sciences for a potential climate change innovation system.

4.2   The War on Cancer and Project Independence

A useful starting point for analysis is to compare the origins of innovation systems for life sciences and alternative energy in the United States. In particular, while the rise of American hegemony in science and engineering after World War II was concentrated in areas such as computing, petrochemicals, and aeronautics (Nelson and Wright 1992; Mowery and Rosenberg 1998), the 1970s saw novel sustained efforts to devote significant and sustained funding to fundamental challenges in both life sciences and energy.

On the one hand, prompted by a combination of public furor over the prevalence of cancer (at that time the second leading cause of death among Americans) and optimism on the part of researchers over recent advances in immunology and oncology, the Nixon administration initiated the War on Cancer. In his 1971 State of the Union Address, Richard Nixon staked out innovation-oriented progress on cancer and the life sciences more generally as a priority:

> I will also ask for an appropriation of an extra $100 million to launch an intensive campaign to find a cure for cancer, and I will ask later for whatever additional funds can effectively be used. The time has come in America when the same kind of concentrated effort that split the atom and took man to the moon should be turned toward conquering this dread disease. Let us make a total national commitment to achieve this goal. (Nixon, State of the Union, January 25, 1971)

Bipartisan support for this initiative led to the 1971 National Cancer Act, which significantly increased the budget for the National Cancer Institute and established the Frederick Cancer Research and Development Center (which would ultimately become an important home both to cancer research and the development of tools, materials, and data infrastructures important to cancer research). The funding and policy commitments initiated by the War on Cancer (and related scientific and technological developments discussed in the next section) were the foundations for a slow but steady growth in the federal commitment to basic life sciences research, grounded, by and large, in advances in molecular biology and genetics. Over time, the NIH ended up being responsible for a disproportionate share of all federal research expenditures on basic research within two decades (Stern 2004).

A similar combination of public concern and technological optimism fueled Project Independence, an innovation-oriented energy independence initiative proposed by the Nixon administration only three years later in response to the 1973 oil crisis. In the 1974 State of the Union, Nixon prioritizes innovation and new technology as a solution for energy independence:

> As we move toward the celebration 2 years from now of the 200th anniversary of this Nation's independence, let us press vigorously on toward the goal I announced last November for Project Independence. Let this be our national goal: At the end of this decade, in the year 1980, the United States will not be dependent on any other country for the energy we need to provide our jobs, to heat our homes, and to keep our trans-

portation moving. To indicate the size of the Government commitment, to spur energy research and development, we plan to spend $10 billion in Federal funds over the next 5 years. That is an enormous amount. But during the same 5 years, private enterprise will be investing as much as $200 billion—and in 10 years, $500 billion—to develop the new resources, the new technology, the new capacity America will require for its energy needs in the 1980's. That is just a measure of the magnitude of the project we are undertaking. (Nixon, State of the Union, January 30, 1974)

Not simply a matter of political rhetoric, both the Nixon and Carter administrations indeed invested significant federal resources in alternative energy initiatives throughout the 1970s, primarily through the Department of Energy projects such as the Clinch River Breeder Reactor and exploratory basic and applied research into synthetic fuels. Indeed, while the level of federal support declined after 1980, there is a case to be made that the foundations for an American alternative energy system began to be established, in a tenuous way, at that time.

In other words, both the life sciences and alternative energy innovation systems were at roughly similar levels of development and scale in the United States in the mid- to late 1970s. However, starting in the late 1970s (and particularly after the beginning of the Reagan administration), there was a dramatic and sharp divergence in the growth and evolution of each of these systems. On the one hand, the alternative energy innovation system was largely dismantled, characterized by scattered and isolated projects. In contrast, the life sciences innovation system embarked on a long period of systematic growth, founded on an orientation toward innovation and a commitment to a step-by-step research process involving complementary investment utilizing both public- and private-sector resources.

4.3   The Life Sciences Innovation System

We now turn to a more systematic analysis of the growth and evolution of the U.S. life sciences innovation system, building on the innovation system literature (Nelson 1993; Lundvall 1992; Mowery and Nelson 1999). In particular, we present a brief narrative history that focuses first on the combination of economic, institutional, and technical conditions that allowed the life sciences innovation system to emerge in the 1970s and 1980s and then consider the key drivers of the growth and evolution of that system over the past two decades.

Our analysis begins by defining what we mean by the life sciences innovation system. Simply put, an innovation system consists of the interrelated and interdependent web of institutions and entities that contribute to the exploration, development, commercialization, and diffusion of new knowledge and technology. The overall productivity of life sciences research and commercialization efforts depend on the structure of the innovation system, including the participation and role of public and private institutions and the nature of the relationships among institutions (and the people within them). Within the U.S. life sciences innovation system, these institutions include, but are not limited to (a) sources of capital and funding such as the NIH, private philanthropy, venture capital, and the investment activities of public companies; (b) sources of research performance including basic university science departments; academic medical centers that combine basic research, clinical research, and patient care, private biotechnology start-up innovators; and the research activities of established pharmaceutical and medical device companies; and (c) sources of commercialization capability including downstream pharmaceutical firms and the activities of supporting sectors such as the clinical research organizations. As we describe in some detail in the following, this system has realized an extraordinary rate of scientific discovery and nascent technological innovation; at the same time, there are concerns about the ability of this system to translate these promising scientific and technical developments into products and tools that are able to overcome regulatory barriers and achieve a high level of diffusion.

The section can be divided into the various "eras" of the life sciences innovation system. Up until the 1970s, the pharmaceutical industry was mostly isolated from the molecular biology and genetics, producing a very different environment than the current life sciences innovation system. In the 1970s, we see a distinct shift in public and private funds for basic research in molecular biology and related fields. From 1980 to 1995, the development of a skilled and specialized workforce combine with the "biotech gold rush" to create an emerging innovation system (though the number of products and therapies that are commercialized remains quite limited). From 1995 onward, the life sciences innovation system has matured as more stable platforms and institutions for cumulative research emerge. Over the past decade, the U.S. life sciences innovation system has served as a dynamic source of commercial applications for new technologies for pharmaceuticals, medical devices, and agricultural biotechnology.

4.3.1   Pre-1970s: The Divide between the Pharmaceutical Industry and Molecular Biology

While the period after World War II saw the rise of the U.S. pharmaceutical industry and the emergence of molecular biology and related disciplines, these two areas of activity remained largely distinct from each other until the early 1970s. By and large, the pharmaceutical industry focused on large-scale, random-screening of drug candidates (Schwartzman 1976; Cockburn, Henderson, and Stern 2000). As a practical matter, this involved work by medicinal chemists who tested thousands of compounds for evidence of a physiological reaction in animal tests (e.g., measuring whether a particular compound lowers the blood pressure of hypertensive rats (Henderson and Cockburn 1994). Large, vertically integrated firms relied on (essentially) serendipity in the earliest stages of the drug development process. As Maxwell and Eckhardt note in their conclusions to their detailed study of thirty-two drug innovation histories, "screening . . . appears to be all but indispensable to the discovery of innovative drugs, having been involved in the discovery of 25 of the 32 case histories covered by us" (Maxwell and Eckhardt (1990, 409). Indeed, during the early 1980s, many researchers expressed strong appreciation for screening methods in the absence of biological theory:

> In some cases it is surprising how well medicinal chemistry can do without knowing the biological system involved. The narcotic analgesics may serve as an example. By means of rather simple screening methods an enormous number of potent and specific analgesics were being and could be developed. (Carlsson 1983, 35)

This brute force approach to innovation was profitable in an environment where the availability of relatively few effective pharmaceutical products, stringent FDA regulation, and broad patent protection resulted in inelastic demand and the ability to charge significant premiums for those drugs that were able to reach the marketplace. In particular, the 1962 Kefauver-Harris FDA amendments (and subsequent regulatory infrastructure developed by the FDA) led both to the systemization of the clinical trial process (including randomized treatments and control groups) and to the erection of significant barriers to entry for those firms that were able to successfully navigate the drug approval system (Thomas 1990). While drug companies did draw on individual scientific findings (through reading journals, etc.) or by hiring skilled graduates, the pharmaceutical industry was primarily engaged in applied industrial research and development activities, and competitive advantage was earned through control over proprietary random-screening techniques and effective clinical trial management (Henderson and Cockburn 1994; Gambardella 1995).

Though emerging at a similar time as the pharmaceutical industry, molecular biology and genetics remained distinct and separate from commercial drug development. Founded in the 1930s, molecular biology focused on fundamental theoretical and empirical questions concerning the function and structure of genetic material. The proposal of a double helix structure for DNA by Watson and Crick in 1953, the most public achievement of molecular biology, ensured the place of molecular biology as among the most elite and basic type of pure science. Prior to the early 1970s, the "distance" between fundamental research in molecular biology and drug development was significant.

In contrast to the chemistry-oriented random-screening approach, molecular biologists sought to address fundamental research questions, even when compared to mainstream biochemistry. While mainstream biochemists focused on characterizing biochemical pathways among eukaryotes (higher organisms [most notably humans or related species]), molecular biologists focused almost exclusively on careful studies of the molecular genetics of prokaryotes (lower organisms lacking cell nuclei; Kenney 1986; Stern 1995). Though these discoveries were important in a scientific sense (resulting in multiple Nobel Prizes), as long as the tools and techniques of molecular biology were limited to lower organisms, the utility of this fundamental research for biopharmaceutical innovation was essentially nil. Moreover, this gap between molecular biology and drug development was not simply a matter of scientific distance: the bulk of the advances in molecular biology were conducted within "classical" academic biology departments that had few if any connections to industry. Academic medical centers such as those that emerged at Stanford were the exception rather than the norm, and the commercialization of any discoveries would have been constrained by perceived limitations on patenting federally funded research and unresolved issues in patent law governing the ability to obtain patents on living organisms or genetically modified biological materials.

4.3.2   1970–1980: The Foundations of Biotechnology

The linkage between pharmaceutical (and medical device) innovation and molecular biology—the foundation of biotechnology and the origins of the life sciences innovation system—can then be traced to a collection of complementary technical, economic, and institutional shifts during the 1970s and early 1980s that bridged the earlier divide: the development of recombinant DNA technology and complementary scientific and technical advances, a significant increase in funding and resources for life sciences research (both public and private, both in the United States and abroad), and a set of policy decisions—such as the 1980 *Diamond v. Chakrabarty* Supreme Court decision and the Bayh-Dole Act—that allowed the assertion of intellectual property rights over innovations based on genetic engineering, even those funded by the public sector.

From a technical perspective, critical advances in technique, instrumentation, and theory overcame many of the barriers that had slowed the application of molecular genetics. The most public of these advances was the gene-splicing technique pioneered by Stanley Cohen and Herbert Boyer in 1973. Along with work by Jackson, Symons, and Berg, the Cohen-Boyer technique gave researchers the ability to manipulate—to change—the genetic code and subsequent protein production of an organism (Johnson 1983). While the Cohen-Boyer was the most public advance, complementary technical advances such as gel electrophoresis and gene synthesis were also achieved during this period. Together, these advances greatly enhanced the potential to exploit molecular biology as a tool or methodology for commercial applications and served as the nascent foundations for biotechnology.

Equally importantly, institutional and policy shifts facilitated the emergence of biotechnology at the university-industry divide. Three policy shifts stand out. First, the Bayh-Dole Act allowed and encouraged researchers at universities to seek patents rights for government-sponsored research (Mowery 2004). The Bayh-Dole Act was meant to increase the benefits to society of public research by incenting inventors to patent and commercialize their work. Around the same time, *Diamond v. Chakrabarty* upheld that genetically engineered organisms were eligible for patent protection, thereby allowing patents on a significant amount of biological sciences research. These decisions had a profound impact on universities, which began to set up technology transfer offices to commercialize their research. The impact on university researchers was equally important, who now saw reduced barriers to patenting and licensing their inventions. These two policy decisions were instrumental in laying the foundations for the dynamic early-stage commercialization environment characteristic of the life sciences innovation system. Third, there was a significant increase in access to private-sector risk capital as the result of the growth of the venture capital model. Not simply a private-sector, "financial-sector" innovation, the growth of venture capital was grounded in policy decisions: the 1979 amendment to the Prudent Man rule allowed pension funds to invest in venture capital, substantially increasing the money available to commercialize technologies. In addition, the NIH emerged as a central player in financing and supporting extramural early-stage research. The scale of funding and its focus on extramural basic research helped to define the NIH's role and sustain its importance in the innovation system. In particular, starting with the War on Cancer, the NIH and related federal life sciences funding grew at a rapid but relatively steady pace, culminating in a funding surge during the late 1990s (figure 4.2).

Extramural NIH funding created a high-level of competition for funds and supported the development of departments in universities focused on the biosciences. Grant-supported training of PhD and postdoctoral students engaging in frontier research helped to create a mobile, knowledge-based workforce that moved between industry and academia.

By the early 1980s, therefore, the stage was set for rapid growth in innovative activity at the interface of academic science and commercial research. In universities, revolutionary discoveries showed the promise of a new frontier for basic science to investigate—and a substantial "payoff" to public funding. In industry, these advances highlighted the viability of making drug discovery and the early stages of the commercialization process more science-intensive. And a new form of organization, science-intensive, venture-backed entrepreneurial firms closely linked to universities and government laboratories had begun to emerge as credible and critical players in the innovation system. This increased focus on basic science set the stage for the development of an entirely new innovation system.

4.3.3   1980–1995: The Emergence of the Life Sciences Innovation System

By the late 1980s and early 1990s, life sciences research had developed a foothold in universities across the United States, and the early stages of what we now refer to as the life sciences innovation system developed. It is useful to focus on three key elements of the system during this period: the development of a skilled and specialized R&D workforce, the introduction of the first generation of biotechnology products, and the emergence of institutions and policies that have reshaped the university-industry interface.

The Development of a Skilled and Specialized R&D Workforce

A distinctive attribute of the molecular biology and genetics communities during the early 1970s was its small size. In key areas, only a small number of researchers and laboratories had the specialized training and tools to take advantage of technologies such as the Cohen-Boyer technique, and this small community was responsible for the bulk of the early activity and advances (Hsu and Lim 2007). Over the 1980s and early 1990s, however, there was a very significant increase in the size and nature of the life sciences workforce, in both the public and private sector (NSF 2008). Figure 4.3 illustrates these trends. Between the early 1970s and today, the number of life science doctorate holders employed in academia has more than doubled (areas such as the physical sciences and engineering have realized a much smaller percentage gain over time), and there was also a significant expansion in the relative number of bachelor-level students who receive a degree in the life sciences fields. Notably, during the first half of the 1990s, areas such as engineering and computer science experienced absolute declines in the number of bachelor's degrees awarded, while life sciences overtook

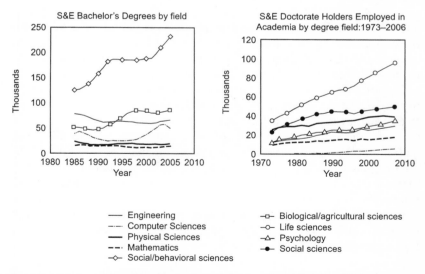

Fig. 4.3   Science and engineering bachelor's degrees by field, 1985–2005, and science and engineering doctorate holders employed in academia by field, 1973–2006
Source: NSF (2008).

engineering as the leading field of study in the "hard sciences and engineering." These are not isolated trends: life sciences consistently graduates the highest number of doctorates of any field, accounts for more than half of all postdoctoral researchers working in universities, and is by far the area with the highest number of academic publications (NSF 2008).

In part, these trends reflect more qualitative institutional shifts: the 1980s were marked by a significant increase in the number and scope of graduate programs in molecular biology, genetics, and related bioscience fields, and many universities established new institutes and departments to take advantage of the new technologies. For example, while the Massachusetts Institute of Technology (MIT) had long maintained a small but high-quality presence in biology and related fields, the establishment and growth of the MIT Whitehead Institute (and related initiatives) during the 1980s changed the character of teaching and research at MIT, with a shift from a dominant emphasis in the physical sciences and engineering toward the development of a large, diverse, and highly productive life sciences faculty. Along with leading research universities, the NIH helped to develop well-defined training and career paths, including the sponsorship of graduate and postdoctoral fellowships, encouraging significant entry by young researchers into the fields that were able to take advantage of the rapidly improving technology.

A Biotechnology "Gold Rush"

The rapid expansion in the scale and scope of biotechnology was driven, at least in part, by the early introduction of a few key "blockbuster" biotechnology drugs that raised exceptionally high expectations for the commercial potential and near-term human health impact of the new technologies. Importantly, the early biotechnology industry was marked by the founding of numerous companies with strong ties to leading university researchers (Zucker, Darby and Brewer 1998), many of which received significant capital from the still-emerging venture capital sector or from new public risk capital programs such as the Small Business Innovation Research (SBIR) program.

Genentech is an illustrative and particularly important example of the types of companies emerging during this period (Hall 1988; McKelvey 1996; Stern 1995). Founded by University of California, San Francisco (UC-SF) researcher Herbert Boyer (of the Cohen-Boyer gene splicing technology) and venture capitalist Bob Swanson in the mid-1970s, Genentech was able to rapidly develop (and patent) two particularly promising applications of the new technologies—human insulin and human growth hormone. These innovations attracted extraordinary interest because they represented a very different type of commercial product (i.e., the genetically engineered production of human proteins) that met an important and unmet human health need. For example, prior to the introduction of Humulin (human insulin) in 1982, the insulin needs of patients suffering from Type I diabetes were met primarily with insulin extracted from pigs or cows (and some human cadavers), with significant clinical limitations (e.g., a significant fraction of patients had adverse reactions to nonhuman insulin products). Interestingly, a distinctive feature of these early Genentech innovations was that the actual commercialization of new products arising from the technology was achieved through cooperative commercialization with established pharmaceutical firms (e.g., Humulin was commercialized in a [contentious] partnership with Eli Lilly, which had long dominated the market for nonhuman insulin products). Most notably, even before *any* products were on the market, Genentech was able to attract significant venture capital funding (including a seed-stage investment by Kleiner Perkins) and a liquidity event for these venture capital investors with an enormously successful initial public offering (IPO) in 1980.

The success and excitement of the Genentech IPO led to the first biotechnology "gold rush," with rapid increases in venture capital (and public equity market) funding of biotechnology during the early 1980s, followed by a "bust" period during the mid-1980s as the number of new biotechnology products stagnated. On the one hand, these investments reflected the belief that the ability of Genentech to identify a few straightforward and important applications of the new technology implied a rapid increase in the number of products that would be commercially viable. In actuality, the ability to exploit the new technology was confined to the relatively small community of researchers that had specialized expertise, and there were only a small number of viable target applications. Indeed, in many areas (such as monoclonal antibodies), multiple companies raced against each other to achieve particular technical milestones (for which patent protection would be available). Ultimately, the combination of a rapid influx of capital and the fact that only a very small number of new products were actually introduced implied that average private-sector returns were low, resulting in a period of investor disillusionment and declines in private-sector funding. This boom-and-bust financing cycle is recurrent, with at least four distinct cycles between 1980 and 2000. In each case, the promise of a new application of biotechnology seems to have resulted in significant overshooting by private-sectors investors, resulting in a highly variable rate of private investment environment over time. Importantly, this variation in private funding was buffeted by the ever-increasing and less-variable level of federal support for life sciences innovation research, including funding specifically directed to start-up innovators through programs such as the SBIR. By the late 1980s, life sciences research funding was about evenly split between federal funding and other sources (mostly private-sector risk capital; figure 4.4).

Emerging Institutions at the University-Industry Interface

The final shift in the life sciences innovation environment during this period was the development and evolution of a set of complex and interde-

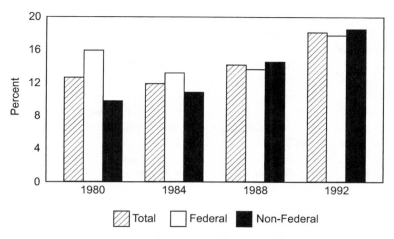

Fig. 4.4   Funding of health R&D as a percentage of total R&D by source
Source: See appendix tables 4-4 and 4-28, NSF (1993).

pendent set of institutions supporting the life sciences innovation system. While a comprehensive accounting of these institutions is beyond the scope of this chapter, it is useful to highlight some key examples, including the development of the modern academic medical center as a locus of research and the establishment of innovation-oriented industry organizations such as the Biotechnology Industry Association. By and large, these new institutions served as extraordinarily effective mechanisms for harnessing the new types of research that were being conducted in the life sciences, and, over time, promoted the development of a new pattern of academic and commercial interactions between universities, start-up innovators, and more-established firms in the emerging innovation system.

Consider the rise of academic medical centers (AMCs; Rosenberg 2008). While academic teaching hospitals had long played an important role in physician training and clinical research, most basic research in molecular biology and genetics was conducted in university biology departments that were largely insulated from practical applications (and were focused on research that that had little near-term practical application). During the late 1950s and 1960s, a small number of universities, such as Stanford and UC-SF, pioneered an alternative approach in which basic life science research disciplines such as molecular biology were established as independent departments within the medical center, supported by a range of programs from NIH and private foundations. In the case of Stanford, the transformation of the medical school to an academic medical center began with the move in 1959 from a San Francisco location to the main Stanford campus and the recruitment by Fred Terman of a range of biochemists and molecular biologists such as Arthur Kornberg and Paul Berg and geneticists such as Joshua Lederberg. Each of these researchers was drawn from a traditional science department. While AMCs such as Stanford were considered oddities at the time of their initial inception, these centers turned out to be extraordinarily productive basic research environments (resulting in numerous Nobel Prizes) that additionally created technologies and tools with significant practical application (including the founding of successful companies adjacent to the Stanford campus that aimed to commercialize these discoveries; Rosenberg 2008). This new organizational model allowed frontier researchers to pursue life sciences research that increasingly took place in "Pasteur's Quadrant," where a single research finding can be a fundamental scientific discovery and serve as the basis for a commercially oriented new technology (Stokes 1997; Murray and Stern 2007).

By the 1980s, there was a significant shift toward the AMC model across leading American universities. To highlight but one notable example, the Harvard University Biology Department had long been a leader in fundamental biological research, under the long-term leadership of James Watson, and leading researchers from that department such as Walter Gilbert had been involved in the early years of the Biotech gold rush as a cofounder of Biogen (Hall 1988; Stern 1995). However, during the 1980s, the locus of a significant fraction of research activity and talent at Harvard shifted toward the Harvard Medical School, with the establishment of new basic research departments. In many cases, these new departments received significant support from industry (along with NIH and foundations), occasionally raising key challenges for the management of the university-industry interface. For example, one of the first discoveries of the newly formed Genetics Lab at Harvard Medical School was the OncoMouse (a mouse genetically engineered to be predisposed to cancer developed by Phil Leder and Tim Stewart), which became the first genetically engineered mammal to receive U.S. patent protection (Murray 2009). Though the discovery was made at Harvard, the funding agreement underlying the research resulted in an exclusive license to DuPont, which enforced its intellectual property (IP) rights aggressively (even threatening enforcement against follow-on academic researchers), resulting in a significant controversy within the life sciences community that was only resolved by an agreement between DuPont and the NIH in 1998 in which DuPont agreed to allow academic researchers free access to the technology. The granting, licensing, enforcement, and NIH settlement regarding the OncoMouse patent was emblematic of the novel challenges that arose as life sciences research increasingly came to have a dual existence on both sides of the university-industry divide (Murray 2009).

At the same time, the nascent biotechnology industry began to build more durable institutional structures that reflected its orientation around innovation and the translation of basic life sciences research. Most notably, while the pharmaceutical industry had long supported an extremely strong industry association (PhRMA) that was largely focused on facilitating a more effective regulation of drug introduction, marketing, and reimbursement, the Biotechnology Industry Organization (BIO) focused on nurturing more effective collaboration between universities and industry and between venture capitalists, start-up innovators, and more-established downstream partners. For example, the annual BIO meeting began to combine a wide range of frontier scientific research presentations alongside panels and discussions of best-practices for intellectual property management, licensing, and effective clinical trial management. By the early 1990s, BIO began to develop specific practices encouraging a "market for technology" for biotechnology tools and discoveries, with explicit disclosure rules and the provision of forums for effective collaboration.

The rise of academic medical centers and the development of a distinctive and innovation-oriented industry association are but two key developments in the institutional framework undergirding the emerging life sciences innovation system during the 1980s and early 1990s. Among other developments, there has been significant entry and growth of specialized suppliers of biomedical materials and tools (including gene sequencers, biomaterials, etc); the development of contract research organizations that can provide expertise in areas such as early-stage clinical trials; and the development of specialized managers, lawyers, and venture capitalists who provide expertise and reputation facilitating more effective transactions in what became an increasingly complex web of relationships between academe, entrepreneurs, and downstream firms.

4.3.4   The Mid-1990s Onward: A Mature Life Sciences Innovation System

By the mid 1990s, the mature structure of the modern life sciences innovation system began to emerge. While there is no single event or marker delineating this more mature system from its earlier incarnation, several events during the mid-1990s altered the character and ultimate scope of the system. First, several enabling platform technologies such as polymerase chain reaction (PCR) became cost-effective across a range of applications, greatly expanding the scope of biotechnology-oriented research and innovation. Second, the institutional shifts from the 1980s transformed the structure of interaction between public and private life science research organizations, resulting in an extraordinarily complex research network structure. Finally, the significant and sustained investment in the system began to pay off—the number of new therapies with their origins in biotechnology increased after a long period of stagnation, and an increasing share of all new drug development began to be grounded in biotechnology and the life sciences innovation system.

Platforms for Cumulative Research and Innovation

While the discoveries of the 1970s and 1980s represented fundamental scientific breakthroughs and offered isolated commercial applications (such as the development of synthetic insulin and human growth hormone (Stern 1995; McKelvey 1996), the growth the life sciences innovation system has ultimately relied not on prototypes but on a cumulative series of complementary technological and scientific breakthroughs. The maturation of the life sciences innovation system was marked by rapid improvements in several enabling technologies that dramatically shifted the productivity and potential scope for life sciences research, including (but not limited to) the development of rapid genetic sequencing methods, the widespread availability of animal research models (such as knock-out mice) that allowed for precise experimentation and inference, and the development of powerful databases such as GenBank (and the data from the Human Genome Project) and ever-more-sophisticated bioinformatics tools to exploit and analyze this data explosion.

Consider the case of PCR, the single most important advance in genetic sequencing technology. Originally developed in the early 1980s by Kary Mullis (a researcher at Cetus Corporation), the use of PCR and the power of genetic sequencing was still quite expensive (perhaps as much as $50 per base pair of a gene) and so was used mostly for small-scale experiments throughout the 1980s. However, PCR was subject to a constant and rapid rate of improvement so that by the mid-1990s, the cost per base pair had been reduced by more than an order of magnitude and has been additionally reduced by two additional orders of magnitude over the last fifteen years (figure 4.5). In other words, while PCR was available as a technology during the 1980s (indeed, Kary Mullis won the Nobel Prize in 1993 for his discovery, and the key patent rights to the technology were purchased for more than $400 million by Roche in the early 1990s), the dramatic improvements in PCR over time, resulting from a long stream of incremental improvements, have transformed the potential applications and scope for research using this technology. While the Human Genome Project—the largest "early" sequencing project using PCR—required sustained investment by thousands of scientists over the entire course of the 1990s, the cost of sequencing an individual human genome is now below $50,000, with strong expectations that individualized genome sequencing will be available as a mass market application within the next three years.

A similar case can be made for each of the other foundational enabling technologies of the life sciences innovation system. While genetically engineered knock-out mice were available in small quantities and a small number of varieties during the 1980s, the late 1990s saw an exponential explosion in the rate of development of specialized research mice (more than 13,000 different mice have now been developed and disclosed in the public scientific

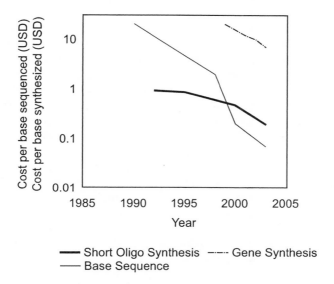

Moore's law—cost per base continues to fall

Fig. 4.5 **Cost per base synthesized and sequenced, highlights the emergence of a cumulative innovation environment driven by General Purpose Technologies (GPTs) including PCR and IT apps**
Source: Estimated from Burrill & Company (2008).

literature). Similarly, while bioinformatics was an exciting basic research area during the 1980s, the development of data infrastructure systems such as GenBank, alongside the dramatic expansions in processing power (and connectivity through the Internet), have transformed the ability to use precise genetic sequencing information in more applied research projects. Overall, the modern life sciences system is marked by a relatively constant rate of cumulative technical progress in a collection of key enabling technologies, allowing for dramatic improvements over time in the scope and potential applications that are able to be addressed by these technologies.

The Life Sciences Innovation Network

The mature life sciences innovation network is also marked by an extraordinarily complex network of structured relationships among research organizations. While a loose network structure of entrepreneurial firms often characterizes industries during their earliest stages that is then followed by a period of consolidation (Utterback 1994), the life sciences innovation network has been marked by sustained and ever-growing interaction and interdependency between university researchers, start-up innovators, and downstream firms engaging in both research and cooperative commercialization. Three features stand out.

First, university research continues to be a central input into the life sciences innovation system. The earliest development of the life sciences innovation system was characterized by the development of start-up innovators in a relatively small range of narrow application areas (such as the commercialization of particular hormones such as insulin) and the increased reliance of a "rational" approach to drug design grounded in biology by pharmaceutical firms (Cockburn and Henderson 1996, 1998). Over time, the potential scope for commercial applications arising from university research has expanded considerably and now covers a wide range of background disciplines and potential application areas. To highlight but one example, developmental biology was long a fundamental area of science with little scope for potential commercial application; with the discovery and characterization of human stem cells in the late 1990s (a key advance in developmental biology itself), fundamental scientific findings in this area became enmeshed in Pasteur's Quadrant, and there was a rapid increase in seeking out formal IP protection for discoveries that were traditionally disclosed exclusively through the scientific literature, the founding of numerous start-up firms seeking to develop these insights for the purposes of licensing and commercialization, and significant new investments by existing biotechnology firms and more established pharmaceutical firms in developing commercially oriented research programs to take advantage of new developments in stem cell science. In other words, rather than the role of basic science receding over time as firms turned their attention toward process improvements and more incremental innovation, the life sciences innovation system has been characterized by a state of "perpetual immaturity" in which university research continues to spawn an ever-wider range of potential avenues for commercial application.

Second, while the traditional pharmaceutical industry had been largely vertically integrated in research, production, and distribution, the modern life sciences innovation system is marked by a diverse range of specialized R&D firms who engage in cooperative development and marketing with more-established downstream players (Gans and Stern 2003). The continuous flow of scientific innovations and the fragmentation of the value chain encourage the biotechnology sector to continuously create new companies. Over time, the biotechnology sector had seen the founding of more than 1,300 companies in the United States and around 5,000 worldwide (Burrill & Company 2008). Although some successful biotechnology companies have ultimately transformed into large firms with a downstream market presence—Genentech and Amgen being prime examples—the sector as a whole is a study in dynamism, with new entrants appearing on the scene every year, and commercialization most often achieved through partnerships and cooperation with more-established companies for development and distribution. Over time, the number of alliances between biotechnology companies and downstream firms has continued to grow, with rapid growth in these arrangements from the early 1990s onward (figure 4.6). An important implication of the presence of a large "market for technology" is that, though the life sciences system is highly innovative, the sector has *not* experienced the widespread creative destruction of established firms in the pharmaceutical industry. Rather than overturning the market power of established companies, university-based entrepreneurship in the life sciences has largely reinforced the market power of the preexisting pharmaceutical industry (Gans and Stern 2003).

Finally, the modern life sciences innovation network is characterized by an extraordinarily high degree of complexity and interdependency and is clustered in a small number of key locations. Not simply composed of bilateral relationships between individual organizations, the life sciences innovation network is highly decentralized and involves multiple linkages between and among different institutions, including universities, start-up firms, established biotechnology companies, pharmaceutical firms, government, and venture capitalists (figure 4.7, drawn from Powell et al. 2005). Both public

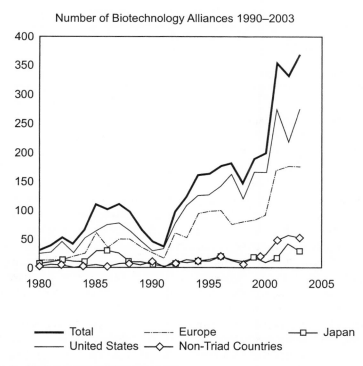

Fig. 4.6   Biotechnology alliances over time
Source: NSF (2006, volume 2, table 4-37).

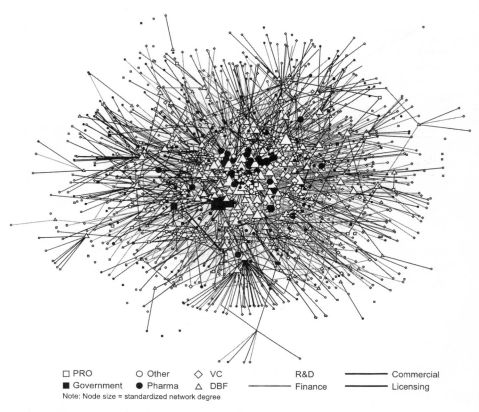

□ PRO   ○ Other   ◇ VC   R&D ─────── Commercial
■ Government   ● Pharma   △ DBF   ─────── Finance   ─────── Licensing
Note: Node size = standardized network degree

Fig. 4.7   The Life Sciences Innovation Network (1998)
Sources: Drawn from Powell et al. (2005).
Note: Node size = standardized network degree.

institutions and private firms play key roles in the network (though there is no one influence that dominates), and the network structure has become considerably more complex and interdependent over time (the 1998 network in figure 4.7 is considerably denser than the equivalent network structure from 1988). Importantly, this highly evolved system is centered in a few key locations (such as the Boston area, the San Francisco Bay Area, and the area around San Diego), and each of these regional clusters is marked by a network with a high level of overlap between public and private research organizations of different sizes and maturity. An important implication of this emergent network structure is that the performance of the system is mostly independent of the actions and strategies of any one organization or firm but depends crucially on the effectiveness of the institutions that support structured knowledge production and transfer between and among research and development organizations.

Biotechnology as the Foundation for New Drug Development

The final key indication of the maturity of the life sciences innovation system from the mid-1990s is simply that the system came to serve as the dominant source of knowledge in new drug development. While the first twenty years of the biotechnology industry were marked by a small number of products, mostly in areas that had not been a traditional focus of the pharmaceutical industry (Lerner 1995), there was a sharp increase in the number of drugs with their origins in biotechnology in the mid-1990s (figure 4.8). By the early 2000s, between 25 to 40 percent of all pharmaceutical sales came from products with their origin in the biotechnology sector, and the vast majority of all new drug candidates were closely linked to biotechnology and the life sciences revolution. Over the last several years, a relatively low rate of new drug approvals, alongside some visible product recalls, has led some to question the efficiency of the life sciences innovation system to effectively serve as the primary knowledge source for downstream pharmaceutical innovation; however, a careful look at the overall pharmaceutical product pipeline (particularly the large number of new therapies working their way through the product approval process) and accounting for the significant improvements that are made over time in existing products through

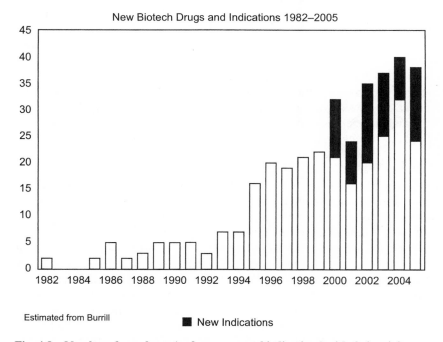

Fig. 4.8   Number of new drugs (and new approved indications) with their origin in biotechnology

Source: Adapted from Burrill & Company (2008).

the discovery of new indication and improved drug delivery suggests that the significant investment in life sciences research over the last thirty years has begun to pay off in terms of a wide range of new and improved therapies with significant human health and welfare benefits (Cockburn 2007).

More generally, this narrative history suggests that the life sciences innovation system has ultimately *replaced* the traditional divide between university science and pharmaceutical innovation with a system that depends on interdependent and collaborative knowledge development spanning both public and private organizations.

4.4   The Drivers of the Life Sciences Innovation System

We now build on this narrative history to identify some of the key drivers of the foundation and growth of the life sciences innovation system over the last thirty years. In particular, we focus on those institutions and environmental conditions that have allowed the system to achieve its high level of dynamism and growth and identify some of the broad innovation policy choices that have impacted the system over time. By characterizing the driving forces underlying the (now mature) system within the life sciences, we are able to turn in the next section to draw out some innovation policy lessons for the (much more nascent) energy and climate change innovation system. We focus in on two broad types of factors: (a) broad supply and demand conditions, and (b) the institutional and strategic environment.

4.4.1   Supply and Demand Conditions

We first highlight the broad supply and demand conditions that have shaped the growth of the system, including the high level and (mostly) stable growth in the level of public funding, the development of a skilled and specialized life sciences workforce, and the presence of a high willingness to pay for breakthrough innovations that address human needs.

The High Level and Growth of Public Funding

While the particular structure and institutional aspects of the life sciences innovation system are undoubtedly important, an extremely important driver of its performance has been the sustained and growing long-term public investment in life sciences research, primarily through the expansion of the NIH. Whereas most other areas of nondefense federal R&D funding have either stagnated or declined (in real terms) since the 1980s, life sciences funding has more than tripled (and nearly quadrupled) in real terms (figure 4.9). Strikingly, the entire increase in the real nondefense R&D budget over the past thirty years can be attributed to increases in funding for life sciences research. Where life sciences research was a relatively minor component of federal R&D spending (less than 25 percent), life sciences research makes has made up the majority of all nondefense R&D funding since 2000.

Three interrelated features of this funding should be emphasized. First,

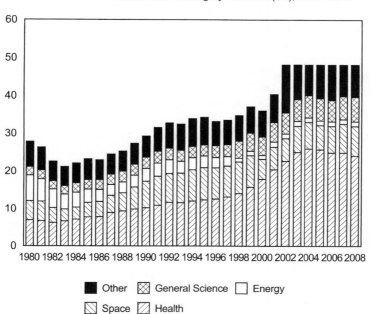

Fig. 4.9   **Trends in nondefense R&D, adapted from the American Association for the Advancement of Science (2009)**
Source: NSF (2008).

from 1980 through the late 1990s, the growth rate in NIH funding experienced very little variability, a sharp contrast to the more volatile private funding environment for biotechnology investment. This steady rate of growth allowed universities and other research organizations to make consistent and coherent long-term planning investments, particularly given that this was an era where the physical capital infrastructure of academic medical centers was considerably expanded. The funding pattern since 1998 has been more variable, with a doubling of the NIH budget in nominal dollars between 1998 and 2003, followed by a flat nominal budget from 2003 through 2008. By 2008, the real declines in funding from 2003 onward implied that the NIH budget had again reached the level it would have reached along the stable growth path that had characterized the 1980 to 1998 period.

At the same time, while the overall NIH budget had a relatively low level of variation in growth (at least until 1998), the funding within the NIH was much more variable. Over time, there have been significant shifts in the particular focus and emphasis of the NIH budget. For example, while funding for the emerging AIDS crisis during the early 1980s was essentially nonexistent (with political resistance for several years), AIDS funding received dramatic increases starting in the mid-1980s and ultimately came to account for a significant share of the overall NIH budget. Similarly, in response to the opportunities afforded by high-throughput sequencing enabled by technologies such as PCR, the NIH, and Congress were able to direct significant increases in funding to genetics and bioinformatics research, both through the peer-reviewed grant system as well as through special initiatives such as the Human Genome Project. While there is, of course, a reasonable level of persistence in the funding for each NIH institute and area on a year-to-year basis, the ability of the NIH and congressional funders to reallocate the NIH budget over time to emerging scientific opportunities and to address particular health care needs has nurtured a system with a high level of adaptability alongside the stability required for infrastructure and human capital investment planning.

Finally, the "surge and retreat" pattern of NIH funding over the last decade offers a cautionary lesson about the impact of public funding on research activity (Freeman and van Reenen 2009). During the five-year doubling of the NIH budget, a large number of universities and other research organizations (including the NIH intramural campus) engaged in significant investment in dedicated physical capital investment in laboratories (often with cost-sharing from the NIH) as well as in the expansion of graduate programs and postdoctoral positions in those areas that were receiving the largest increases in NIH funding (such as genomics). These physical and human capital investments occur over far more than a five-year window (particularly in terms of exploiting the investment in terms of research activity), and many of these investments were made under the expectation that the NIH would continue to grow after the 2003 period, albeit at a slower rate. The subsequent reduction in the size of the NIH budget thus resulted in systematic distortions in funding and research activity that likely have reduced research productivity. The unanticipated increased in the share of costs for physical capital investments falling on universities significantly reduced the ability of universities to provide funding for research and graduate student support. More important, the reduction in real terms of the NIH budget has essentially created a "budget squeeze" that is particularly salient for the generation of young investigators that were trained during the boom period. In particular, the surge-and-retreat pattern had the consequence of inducing a significant increase in the number of tenure-track junior faculty at precisely the time when the NIH grant pool for new investigators became far more limited (Freeman and van Reenen 2009). While significant excess returns to the endowment at leading universities mitigated some of these effects (at least at those schools with large endowments that performed well), an emerging body of evidence suggests that the high level of variability in the aggregate NIH budget over the past decade has likely resulted in a less-productive innovation system and distorted the incentives and career dynamics of an entire research generation.

Slow and Steady Growth of a Skilled and Specialized R&D Workforce

Both the overall NIH funding regime as well as the organization of leading university program has nurtured the development of a skilled and specialized R&D workforce. During the 1980s, numerous universities expanded
their graduate training programs to adapt to new technologies through the
creation of new disciplines, including bioinformatics, genetics, and bioen-
gineering. The slow and steady growth of this workforce, and the interdependencies between public-sector and private-sector research had distinct
implications for the structure and performance of the life sciences innovation system.

First, the large size of the emerging discipline meant that individual researchers could become highly specialized and collaborate with other researchers (as coauthors) on particular projects (Wuchty, Jones, and Uzzi
2007). The combination of specialization and opportunities for collaboration cannot be overstated. In the absence of norms and institutions that
encouraged widespread collaboration, individual scientists would have to
master a much wider range of skills and knowledge in order to work on
any one project; this requirement for breadth would come at the expense of
depth, lowering research productivity (Jones 2008).

At the same time, this highly specialized graduate training could result
in a range of alternative employment opportunities, including a traditional
tenure-track position within a biology department; a research position
within an academic medical center that might involve a heavier reliance
on sponsored research activities (as opposed to the "freedom" of a pure
biology department); or employment with a biotechnology, pharmaceutical,
or medical device firm where basic research responsibilities would be complemented with direct concerns about the commercialization of discoveries.
Moreover, the highly interdependent nature of the life sciences innovation
network has the consequence that a period of employment in the private
sector need not come at the expense of returning to public-sector scientific employment in the future. Researchers on both sides of the university-
industry divide publish heavily in the scientific literature, collaborate with
each other on projects, and engage in more structured interactions in the
context of commercialization (Cockburn and Henderson 1998). Indeed, the
career concern incentives to participate in this system on an ongoing basis
may be an explanation for why scientists seeking private-sector employment
are willing to accept lower-wage income for jobs that permit some freedom
in research project choice and permit ongoing publication in the open scientific literature (Stern 2004).

In other words, a distinctive feature of the life sciences innovation system
is that individual researchers are (by and large) able to make very specific
human capital investments early in their career, are able to realize the benefits of those investments by obtaining diverse types of employment that

are nonetheless closely related to their human capital investments, and are able to collaborate with researchers (across organizational boundaries) with complementary human capital over the course of their career.

Significant Financial Rewards for Innovative Clinical Breakthroughs

A third driver of the growth and evolution of the life sciences innovation system is the prospect for significant financial rewards for successful innovation. Most consumers have a very high intrinsic willingness to pay for drugs (as for other medical innovations), particularly when the alternative to drug therapy involves suffering from a painful, debilitating, or even deadly medical conditions. While multiple therapies may exist to treat a condition, most innovative products are strongly differentiated from each other in terms of pharmacological or therapeutic characteristics (Ellison et al. 1997; Stern 1996).

Several interrelated features shape the product market rewards available for innovative products. Consider the environment within the United States (by far the single largest market). First, in terms of the demand for a therapy, pharmaceutical demand is determined largely through physician prescribing choices (who are, therefore, insulated from the pricing impact of their decisions), and a significant portion of patients receive some form of public or private insurance coverage for pharmaceutical purchases. While insurance companies constrain both physician discretion and patient insurance through the use of formularies, it is nonetheless the case that the key decision makers for many pharmaceutical purchases are insulated from the full cost impact of those decisions and, in any case, have a high intrinsic willingness to pay. At the same time, the FDA regulatory framework alongside broad and enforceable patents implies that the substitution choices for an innovative new product are often limited during the time of FDA exclusivity. In particular, the Hatch-Waxman Act of 1984 provided the modern regulatory framework that ensured a reasonable period of exclusivity (for the marketing and sales of a compound) for innovating firms after FDA approval and encouraged the entry of generic firms to promote competition after that exclusivity period had expired (Grabowski and Vernon 1996). The Hatch-Waxman Act was a significant policy success, simultaneously sharpening the incentives for breakthrough innovation while ensuring diffusion and low-cost access after patent expiration. Similarly, for conditions in which there are only a small number of patients (and so the incentives to innovate may not be sufficient), the Orphan Drug Act provides a less-costly path toward regulatory approval and an enhanced exclusivity period. Many biotechnology firms have indeed targeted their efforts at markets covered under the Orphan Drug Act, both to take advantage of the favorable regulatory framework and because many of underlying disease conditions are particularly well-suited to therapies using the tools of biotechnology. Finally, while public payers such as Medicare are the single largest payers within the market, current policy prohibits explicit price controls. Within this framework, for innovative therapies addressing significant health and welfare challenges, demand is highly inelastic, and innovators are in many cases able to charge high prices (particularly compared to marginal cost) and so realize very significant margins during the time of FDA exclusivity. Once the patent expires, drugs are subject to very rapid and effective imitation by a now-mature and effective generic sector.

Of course, the United States is not the only product market, and biopharmaceutical firms are able to realize returns on their innovations on a global basis. While the incomplete globalization of pharmaceutical products is something of a puzzle (Kyle 2007), the opportunities of a global market are nonetheless considerable. While most countries outside the United States impose some form of price controls (and other institutions regarding insurance and generic licensing also vary), most countries outside the United States provide a significant price premium for truly innovative products or those that address a significant condition for which there is no substitute. With that said, particularly for the experimental therapies emerging from the life sciences innovation system, most biotechnology companies (and their commercialization partners) emphasize opportunities for drug development and introduction into the United States, while at the same time seeking to build a global presence.

Finally, it is important to emphasize that the rewards for innovation are highly skewed, even for those products that are able to navigate through the regulatory system. A small number of "blockbuster" products realize very high sales, with the top 100 products accounting for about one-third of all global revenue, and nearly two-thirds of drugs do not generate sufficient market returns to recoup their development costs (Grabowski and Vernon 1996).

Overall, the combination of intense demand on a global basis and limited competition during a period of exclusivity provide powerful incentives for innovation (Thomas 2004). The "pull" of high margins and sales volumes for successful products and the "push" of intense generic competition work together to generate high returns for successful commercialization of biomedical research in the form of significantly improved products (Finkelstein 2004; Acemoglu and Lin 2004; Scherer 2001). In other words, even though there is a significant level of public research for the U.S. life sciences innovation system, the incentives provided by commercialization provide an equally large incentive, resulting in the (roughly) balanced level of total research expenditure between the public and private sector (figure 4.4).

4.4.2   The Institutional and Strategic Environment

Beyond the broad supply and demand factors already highlighted, the growth and evolution of the life sciences innovation system have been powerfully shaped by the underlying institutional and strategic environment, including a transparent and competitive peer-reviewed grant system grounded in the norms of open science, the availability of formal intellectual property rights to protect innovations and provide opportunities for cooperative commercialization, and the presence of innovation-oriented competition along multiple dimensions and domains.

Peer Review and the Norms of Open Science

By definition, the life sciences innovation system is encompassed with the domain of open science (Merton 1973; Dasgupta and David 1994; Stern 2004). The institutions of open science are subtle and interrelated but are ultimately grounded in three distinctive features: academic freedom, the priority-based reward system, and the freedom to collaborate. By and large, life science researchers in public-sector institutions are free to choose their own research agendas and are given broad latitude in how to approach a particular scientific research question (subject to ethical requirements such as human subjects regulation). By giving researchers freedom to choose their own agenda, a more-diverse range of questions and experiments are undertaken (Aghion, David, et al. 2009), with the potential for significant surprises from "unexpected" directions. At the same time, the priority-based reward system gives researchers credit for the prompt and full disclosure of their discoveries (usually in academic journals, but sometimes in other outlets such as databases like GenBank), accomplishing several interrelated objectives. The priority-based reward system complements academic freedom (researchers have incentives to come up with their own solution to problems that others within their field find interesting), encourages prompt disclosure (as one does *not* get credit if someone else discloses the discovery before you publish), and provides a transparent means for access by future scientists to the body of knowledge in a particular area. As an economic institution, open science encourages a high rate of cumulative knowledge production, and, importantly, offers no enduring monopoly rights over the *use* of that knowledge by future researchers (Dasgupta and David 1994; Aghion, Murray, et al. 2009).

Federal policy toward the life sciences builds upon and reinforces the norms of open science in fundamental ways. While the allocation of public funding for biomedical research, like all federal expenditures, is driven to some degree by the political priorities reflected in appropriations bills, to a great extent the allocation of public funding for biomedical research to specific projects and investigators has been controlled by peer-review of investigator-initiated projects and has responded as much to supply of interesting ideas as it has to demand for solutions to health problems. In other words, the NIH peer-reviewed grant system is embedded within a system that encourages academic freedom and disclosure and reinforces those institutions by providing ongoing incentives for participation and appropriate scientific behavior (e.g., prompt disclosure, a high level of ethical conduct, etc.). While there are, of course, exceptions to these norms (e.g., occasional instances of outright scientific fraud, and, almost more troubling, a more pervasive pattern of limited data withholding in some scientific communities), perhaps the most striking feature of the life science innovation system is how rarely such exceptions occur; most scientists place a high degree of weight on the maintenance of their reputations and behave in ways that protect those reputations and promote the transparency and priority rules for scientific research.

With relatively few exceptions, government-sponsored research has rarely taken the form of Manhattan Project initiatives. Rather, the progress of science has been largely driven by a robust and independent scientific community focused on the intellectual merit and novelty of investigator-generated research proposals, balanced by input concerning social or governmental priorities. In the few cases where the life sciences innovation system has focused on a "big science" project, such as the Human Genome Project, the impetus has often come from the recognition of the need for such a project from the scientific community itself.

Intellectual Property as an Incentive Device and an Enabler of a Market for Technology

The nature of IP, specifically patents, has played a very important role in driving innovation in the sector. Several features have been particularly salient. First, in contrast to other sectors, in biopharmaceuticals, the patent system appears to be working relatively well. In other technologies and industry sectors, patents have become highly controversial, with many economists increasingly skeptical that the patent system is actually promoting technical change. Critics argue that poor standards of examination, growth of patent thickets, and increasingly sophisticated strategic use of IP are raising costs imposed by patents (Jaffe and Lerner 2004). But in biopharmaceuticals, these problems have, at least historically, been much less severe.

To a great degree, this is a function of the nature of technical knowledge and of pharmaceutical products. As products, drugs are normally a single molecule, or simple mixture of molecules, not complex devices comprising hundreds or thousands of distinct patented (or patentable) components. At the same time, innovation has taken place largely in the realm of the "chemical arts," a highly systematized and codified domain of knowledge within which widely accepted conventions for nomenclature and technical practice, extensive professional training of participants, a large an exhaustively indexed scientific literature, and a long tradition of patenting make it straightforward to establish the novelty and patentability of new inventions. Patent rights over new molecules are, therefore, generally straightforward to obtain, to delineate, and to defend. Compared to other technologies, infringement of patents is generally easy to observe (and to establish in court) and "freedom to operate," that is, the absence of others' IP that could be used to hold up new products is relatively easy to establish.

Because most pharmaceutical products are also relatively easy to imitate, patents play a critical role in allowing innovators to appropriate returns from R&D. This creates a strong incentive for innovators to submit high quality applications. At the same time, the nature of product market regulation and competition creates extraordinary rewards from invalidating or inventing around a single patent, and competitors have a similarly powerful incentive to weed out "bad" patents.

Industrial R&D is unusually tightly linked to academic science, traditionally driven by "Mertonian" norms and incentives, which are generally thought to be antithetical to exclusion-based intellectual property rights. Thus, as patenting in this sector has increasingly moved upstream from the product market and into the domain of basic science, the specter of a "biomedical anticommons" has been raised. Concerns have been widely expressed about the inappropriate scope or blocking power of patents on fundamental physiological processes or genetic sequence information, of decreased knowledge sharing among scientists, or the potential for transactional "gridlock" as products rely increasingly the integration or combination of many pieces of independently owned IP. But evidence of a serious negative impact on innovation is thus far relatively weak. (Walsh et al. 2003; Murray and Stern 2007). The prominence and scale of open science also serves to vigorously delineate the public domain in biomedical science. And although patenting of research tools, drug targets, genes, and fundamental biological and chemical processes remains controversial, these property rights are the foundation of a very active "market for technology," which has promoted extensive disintegration of R&D organizations, a high degree of specialization, and injected market pricing and entrepreneurial energy into the research process.

Second, in this sector, patents work alongside other mechanisms to generate extraordinarily high rewards to innovative individuals across multiple domains. Successful researchers can expect to gain both high social and professional status (awards, peer recognition, influence), significant intrinsic rewards from their work (curing disease), and substantial financial benefits (tenure, salaries, equity).

Third, patent protection and incentives are closely tied (in the United States, at least) to a regulatory system that controls access to the product market. The Hatch-Waxman framework protects innovators through patent term extensions, data exclusivity provisions, and automatic thirty-month stays preventing the FDA from approving the sale of allegedly infringing generic versions of a drug. But at the same time it provides incentives for imitators to challenge weak patents (180 days of exclusive generic status for the first entrant that successfully challenges and incumbent's patent), forces clarification of the patent status of a drug (innovators must list relevant patents), and substantially lowers entrants' costs of obtaining FDA approval for their product by allowing them to use the innovator's health and safety data in preparing an Abbreviated New Drug Application (ANDA). Together with state laws requiring automatic brand-generic substitution and widespread use of tiered copayment schemes and formularies by insurers and third-party payers, these provisions facilitate deep and rapid loss of sales once patent protection is lost. Thus, the "carrot" of effective and workable IP combined with the "stick" of intense postpatent competition create powerful incentives to compete through innovation.

Competition across Multiple Dimensions and Domains

A distinctive feature of the life sciences innovation system, and one of the key drivers of its innovation performance, is the pervasiveness of competition at multiple levels of analysis. This competition occurs in many dimensions and throughout the value chain. First, consider the "upstream" domain of basic research and scientific discovery. Here, individual researchers compete for priority in discovery and for reputation within their peer community. A rich literature in the economics and sociology of science has identified the key elements of the governance and reward systems in "Open Science" (Merton 1973; Dasgupta and David 1994). While the social norms of these communities place a premium on collegiality, cooperation, and sharing, the highly skewed distribution of rewards (grants, promotion, status, power, "parking") and relatively free entry into the academic labor market results in vigorous competition among researchers (Dasgupta and David 1994; Nelson and Rosenberg 1994). At higher levels of organization, reputation-based competition is equally vigorous between research groups (labs) and between universities and other nonprofit research institutions to attract resources and talent. Notwithstanding the "Matthew Effect" and other competitive dynamics that tend to cumulatively reinforce small performance differences, academic research activity is remarkably "atomistic" and fragmented, with several hundred institutions playing a significant role. At an even higher level of aggregation, states, regions, and countries can be understood as competing in the domain of basic research through policy choices and provision of physical and institutional infrastructure intended to attract human capital and investment. Consider competition between California and Massachusetts in the era of restricted federal funding for stem research or initiatives such as the Singapore "Biopolis."

Another important arena is the "market for technology" (Arora, Fosfuri, and Gambardella 2001; Gans and Stern 2002, 2003; Cockburn 2004). This trade in "technology" (frequently candidate molecules, but also research tools and data) across institutional boundaries has become a critical aspect of the innovation system. Licensing deals, collaborative research, and corporate mergers and acquisitions (M&A) are some of the most salient features of the biomedical landscape. Many institutions compete actively on the supply and demand sides of this market, with universities and academic medical centers, government labs, biotechnology companies and some parts of "Big Pharma" companies competing in the supply of out-licensed technology, and downstream specialists in commercialization (principally Big Pharma, but increasingly "Big Bio") competing to acquire the most promising discoveries to fill development pipelines and maximize utilization of their manufacturing, distribution, and marketing capacity.[2] A distinct set of institutions and marketplaces are emerging to facilitate and govern this trade, such as university technology licensing offices (TLOs) and industry-wide gatherings such as the BIO conference.

Last, there is competition in the product market. As described in the preceding, the nature of demand and the regulatory framework controlling access to the market have focused commercialization activity on product-oriented innovation. The increasing role of generics and the maturation of that part of the industry focused on therapeutic proteins, which are very costly to produce, may result in greater emphasis on process innovation and lowering production costs. But for the most part, commercialization activity is directed toward developing novel premium priced molecules and generates intense "Schumpeterian" dynamic competition between innovative products. Lichtenberg and Philipson (2002) show that a typical drug is launched in the face of twenty-five existing molecules in its therapeutic classes and faces entry by a further seven to ten new molecules introduced during the time it is still patent-protected. As with the basic research sector, competition is generally fragmented and atomistic. Notwithstanding ongoing consolidation among Big Pharma companies, relatively few therapeutic categories are dominated by a single producer.

This complex, multifaceted competition has a powerful influence on the rate and direction of innovation within this sector. Several aspects are worth noting. First, with the singular exception of generic manufacturing, competition for resources throughout the system is consistently oriented around innovation, priority, and the creation of new knowledge. Second, compared to other sectors, the nature of competition and of innovations that results is highly transparent. There is a pervasive culture of codification and disclosure of new knowledge through scientific publication and patenting. This supports a research environment that is strongly cumulative and highly efficient in the sense of avoiding duplication. Third, competition and experimentation thrive in the absence of a single bottleneck, dominant platform, or monopoly player. Ironically, it is the industry where property rights over innovation are likely strongest that we see perhaps the highest sustained levels of innovation-oriented competition in an unconcentrated market. Simply put, there is no Microsoft.

2. By "Big Pharma," we mean the very large companies that have dominated the industry since the 1970s, historically focused on small molecule chemistry-based drugs, and fully vertically integrated from drug discovery through to manufacturing and marketing. "Big Bio" refers to the small set of companies focused on large-molecule technologies that have brought successful products to market and have substantial manufacturing and marketing capabilities.

One particularly interesting, and important, aspect of competition has been institutional and economic experimentation. Though much innovative activity continues to take place within large, stable organizations, this sector has seen dramatic industrial restructuring and the emergence of new and interesting organizational forms such as AMCs that combine bench research with clinical practice (Rosenberg 2008), the "just off-campus" biotechnology firms founded by academic "star scientists"; nonacademic, not-for-profit research institutions such as the Jackson Laboratories; contract research organizations (CROs); specialized venture capital firms; large-scale, hands-on funding by philanthropies such as the Wellcome Trust and the Gates Foundation; patient advocacy groups; and hybrid private-public entities such as OneWorld Health. Just as there is no monopoly on scientific discovery or entrepreneurship, there is no monopoly on the institutional approach to encouraging an effective life sciences innovation environment. This diversity encourages competition and experimentation over time facilitates systematic learning regarding the "science of science management."

4.5   Lessons for Alternative Energy and Climate Change Innovation

The principal contention of this chapter is that the relative dynamism and performance of the life sciences over the past twenty years does not simply reflect scientific and technological opportunity. Instead, the performance and character of the life sciences innovation system is grounded in the microeconomic and institutional environment, and, by and large, these factors have been conducive to significant and pervasive growth. Moreover, the life sciences innovation environment did not arise by chance—rather, it reflects a long history of public policy choices and an institutional framework that provides a robust supply of innovation inputs (money and people), the potential for significant rewards from breakthrough innovation, an engagement with the norms of open science that nonetheless allows for effective IP protection, and opportunities for competition along multiple dimensions.

There is, of course, an ongoing and vibrant policy debate about how best to ensure and enhance the vitality of the life sciences innovation system. In particular, in light of the relatively low rate of approval of new drugs over the past several years, there are increased concerns about whether the system has sufficient capacity and incentive structures to translate scientific discovery into clinical applications that have a significant human health benefit. At the same time, there is a broader health care reform debate that focuses in large part on the ability to reduce the rate of growth of medical costs over the long term; and this broad policy objective may conflict with the historical commitment to reward breakthrough innovations though significant price premiums during the time of FDA or patent exclusivity.

Despite these ongoing challenges, we argue that the history of the life sciences offers an instructive lesson regarding the growth and evolution of science-based innovation systems. It is, of course, important to recognize that there are fundamental differences between the environment for innovation in the life sciences and climate change. One obvious difference is in the mechanisms for rewarding innovators by allowing them to share in the social surplus associated with new technologies. Intellectual property, the regulatory regime, and the payment system for health care mean that innovators in biopharmaceuticals are able to capture a substantial portion of the value generated by new drugs, at least in the short term. The same cannot be said for climate change technologies: absent an effective and durable set of policies for generating large private rewards to innovators in climate change technologies (such as the economic benefits from lowering emissions under a regime with carbon pricing, whether in the form of a carbon tax or a cap-and-trade system), incentives for innovation and commercialization will be muted relative to those in the life sciences. In addition, while the life sciences revolution was spurred by the development of particular discoveries and technologies from the early 1970s that pointed very clearly to feasible biotechnology applications, such as large-scale production of therapeutic human proteins or monoclonal antibodies, the impetus for a climate change innovation system is grounded in a specific social challenge that has so far resisted "easy" technological solutions, and there is no one technological paradigm on which to focus.

But despite these differences, it is nonetheless true that while the innovation system in both life sciences and alternative energy innovation have somewhat similar origins in the early 1970s, investments and progress in alternative energy innovation have been elusive, and the life sciences innovation system has come to occupy a dominant role in nondefense public funding and a leading role in the overall American innovation system. Why is this the case?

4.5.1   Lessons from the Life Science Innovation System

First, the returns to life sciences investments by both private and public entities have taken *decades* to pay off and are only now coming to occupy a central role in the delivery of new therapeutics. These payoffs reflect the slow-and-steady evolution of a complex set of institutions and technologies, supported by sustained and relatively stable public investments. In contrast to a Manhattan Project approach in which a single burst of focused investment yields a single technological fix, the life sciences innovation system has been characterized by steady and cumulative progress over time and the development of complementary platform technologies. Indeed, a single R&D surge with no follow-through might actually be counterproductive in terms of long-term technical progress, as specialized investments are undertaken during the boom period, resulting in significant distortions as funding is cut back. The experience of the life sciences sector further suggests that stable and long-term public funding of research is particularly important in environments that are likely to be characterized by a high degree of interaction between public and private research funders. The private funding of innovation-intensive sectors is notoriously fickle, and an often overlooked benefit of a stable pattern of public funding is the ability to buffer the variability of private-sector investment. It seems likely, therefore, that any systematic and robust effort toward alternative energy will be most effective if it is grounded in a long-term commitment involving the development of specialized human capital and the evolution of institutions that allow for effective public-private interaction. It is often remarked that Rome wasn't built in a day—from an economic and innovation policy perspective, the lesson is that the design principles and technologies that comprised ancient Rome took centuries to develop. Thus, while the social challenges and time constraints presented by climate change have led to considerable pressure to engage in an accelerated innovation process, effective long-term solutions to climate change and energy requirements are more likely to be grounded in a systematic and long-term R&D commitment.

Second, life science innovation has been driven by investigator-initiated and peer-reviewed science rather than a command-and-control approach. Even when particular public health priorities have emerged (as in the case of AIDS), the source of the ultimate solutions have been grounded in the open scientific community and are dependent on the exercise of intellectual freedom and scientific openness and opportunities for experimentation and diversity at the level of individual researchers and institutions. Obviously, product market incentives steer resource allocation in commercial science, but these have been complemented and counterbalanced by the robustness and scale of "blue sky" research—with long-run benefits to all. An environment that encourages academic freedom and entrepreneurship will, of course, be less focused on specific, immediate problems than a command-and-control approach, as different researchers and firms experiment with alternative approaches based on individual perceptions and beliefs. Of course, most of these ideas and approaches will fail, and, even in the life sciences, there is significant pressure to reduce the rate of failure through a more top-down approach. However, the history of the life sciences suggests that attempts to significantly reduce the freedom of investigators and entrepreneurs rarely results in important breakthroughs precisely because it reduces the diversity of experimentation. Efforts to manage the direction of research in a centralized manner, rather than through a peer-reviewed, investigator-initiated system for setting research priorities, therefore, seem unlikely to provide a cumulative stream of innovation addressing the need to mitigate global climate change.

Third, competition is intense and pervasive throughout the value chain in life sciences. Despite consolidation among Big Pharma companies, the product market is relatively fragmented and driven by Schumpeterian competition to introduce new molecules, combined (in due course) with price/cost competition within existing molecules from generics. In the "market for technology," thousands of smaller science-based entrepreneurs compete for capital, human resources, and opportunities to license to or collaborate with downstream partners. In the publicly funded sector, Darwinian competition for resources prevails between many hundreds of institutions and thousands of principal investigators (PIs). Whether directed toward Nobel Prizes or blockbuster drugs, this competition is focused on novelty and priority rather than preemption of scarce resources or control of distribution. Competition at multiple levels and multiple domains enhances the level of experimentation within an emerging innovation system and mitigates the potential for holdup and rent-seeking. Lack of competition within innovation does not simply engender the traditional static losses of monopoly pricing, but also reduces the level of diversity and experimentation of the research community itself, with negative consequences for the rate of cumulative technical progress and the productivity of the resources employed.

Fourth, though providing significant rewards for innovators and firms who effectively commercialize important innovations is extremely important, the dynamism of the life sciences depends on more than simply setting the right "price" for innovative therapies. The life sciences, almost by definition, rely extensively on the norms of open science and engagement with the scientific community, including university researchers. While scientists are naturally motivated by potential financial returns, they are *also* motivated by innate curiosity and the potential for recognition (in the form of prestige, positions, and awards). In some cases, placing an extreme emphasis on financial incentives may actually reduce engagement and participation by the scientific community, particularly the exploitation of IP rights or financial incentives are perceived to be getting in the way of "good science." A fundamental feature of open science is indeed its openness—the (mostly) prompt disclosure of new discoveries through scientific publication (and perhaps complemented by patent filings) and the development of open-access institutions and infrastructure (such as GenBank or the Jackson Laboratories) that enhance the productivity of all scientists. While individual scientists may engage in strategic behavior (e.g., by only partially disclosing their work in order to limit rivals' access to knowledge), the policy choices of the NIH and the governance of academic societies and universities have created powerful norms that enhance the transparency of the knowledge accumulation process. By facilitating access to prior discoveries and providing incentives for the disclosure of new discoveries, open science serves as powerful institutional framework for step-by-step scientific and technical progress.

Fifth, whether by accident or design, the interaction between the patent system, the FDA regulatory process, and the payer environment provide large and very visible incentives for breakthrough innovation. The combination of a high willingness to pay for products (combined with insurance, which insulates purchasers from the marginal price) and the Hatch-Waxman regulatory framework provide firms incentives for develop blockbuster therapies (particularly focused on the largest markets) and to develop a stream of innovations over time (as the monopolies generated by the system are transitory). The Hatch-Waxman framework helps to ensure that innovators are able to recoup the costs of the drug discovery and development process, while also enhancing the diffusion of valuable therapies at low prices after patent expiration. Note that the relative strength of the patent system in this environment not only enhances incentives for innovation but also encourages the development of a market for technology so that therapies discovered within academia or by start-up firms can be brought to market by leveraging the complementary assets and resources of firms more experienced at navigating the FDA process (reducing the time to approval) and with a larger presence in the product market (enhancing diffusion). Importantly, it seems that the discovery, development, and diffusion of new drugs is more efficient when there is significant overlap between the types of innovations that can be covered by a patent and the scope of exclusivity offered by FDA regulation. While the specific institutional framework for drug development is unlikely to be an effective model or analogy for climate change technology, the experience of life sciences innovation suggests that regulation governing product market access can play a crucial role in shaping innovation incentives and that product market regulation and IP rights policy can be powerful complements to one another.

Sixth, the wide spectrum of organizations and institutions that make up the life sciences sector have demonstrated considerable flexibility and adaptation to an evolving environment. New organizational forms, such as the "Dedicated Biotech Firm"—science-based entrepreneurial enterprises that engage closely with academic institutions and star scientists and operate very far from the final product market—have emerged. At the same time, incumbent firms have, by and large, not shown the structural rigidity and organizational inertia that appear to have been so costly in other sectors such as the U.S. automobile industry. Rather than engage in systematic resistance to the emerging biological sciences, established firms accommodated and adjusted to entry of new biotech firms (and the expansion and greater engagement of universities and AMCs), ultimately becoming enmeshed in a web of collaborative institutions and partnerships with both public and private entities. While the management of such research networks is daunting, they represent an effective approach to commercialization grounded in cooperative relationships across a wide variety of institutions. These collaborations allow for specialization in the division of innovative labor and commercialization activity and the exploitation of distinctive complementarities between public science, science-based entrepreneurship, and traditional pharmaceutical companies. Importantly, these relationships are sustained by balancing the norms of open science with effective IP rights.

Our final lesson concerns the nature of life sciences innovation. While many discussions of potential technological solutions for climate change effectively envision a single discrete "quantum leap" that offers a cost-effective substitute to carbon-based energy sources or an ex post mitigation scheme for removing emitted carbon from the environment, the history of life sciences innovation is that most "breakthrough" technologies depend on a long, drawn-out process of cumulative step-by-step innovation, which ultimately delivers significant results after decades of sustained investment and development. As in many other technologies, in the life sciences, embryonic prototypes often provide little indication as to the ultimate social impact of a given technology. To take but one striking example, sustained reductions in the costs of genetic sequencing have ultimately enhanced access to that technology and have facilitated applications across diverse application areas from criminology to public health (characterizing different flu viruses) to personalized medicine. Rather than considering whether there is a single magic bullet for climate change (and offering a single large prize for success), the experience of the life sciences suggests that sustained investments in general-purpose platform technologies and support for diversity, experimentation, and competition across a wide range of organizations and technologies are more likely to result in a stream of powerful innovations to address pressing social and human challenges.

References

Acemoglu, Daron, and Joshua Linn. 2004. Market size in innovation: Theory and evidence from the pharmaceutical industry. *Quarterly Journal of Economics* 119 (3): 1049–90.

Aghion, Philippe, Paul A. David, and Dominique Foray. 2009. Science, technology and innovation for economic growth: Linking policy research and practice in "STIG systems." *Research Policy* 38 (4): 681–93.

Aghion, Philippe, Fiona E. Murray, Mathias Dewatripont, Julian Kolev, and Scott Stern. 2009. Mice and academics: Examining the effect of openness on innovation. NBER Working Paper no. 14189. Cambridge, MA: National Bureau of Economic Research.

Arora, Ashish, Andrea Fosfuri, and Alfonso Gambardella. 2001. *Markets for technology: Economics of innovation and corporate strategy.* Cambridge, MA: MIT Press.

Burrill & Company. 2008. *Biotech 2008.* Annual Biotechnology Industry Report. http://www.burrillandco.com/bio/biotech_book.

Carlsson, Arvid. 1983. The role of basic biomedical research in new drug development. In *Decision-making in drug research,* ed. Franz Gross, 35–42. New York: Raven.

Cockburn, Iain M. 2004. The changing structure of the pharmaceutical industry. *Health Affairs* 23 (1): 10–22.

———. 2007. Is the pharmaceutical industry in a productivity crisis? In *Innovation policy and the economy.* Vol. 7, ed. Adam B. Jaffe, Josh Lerner, and Scott Stern, 1–32. Cambridge, MA: MIT Press.

Cockburn, Iain M., and Rebecca M. Henderson. 1996. Public-private interaction in pharmaceutical research. *Proceedings of the National Academy of Sciences* 93 (23): 12725–730.

———. 1998. Absorptive capacity, coauthoring behavior, and the organization of research in drug discovery. *Journal of Industrial Economics* 46 (2): 157–82.

Cockburn, Iain M., Rebecca M. Henderson, and Scott Stern. 2000. Untangling the origins of competitive advantage. *Strategic Management Journal* 21 (10–11): 1123–45.

Cortright, Joseph, and Heike Mayer. 2002. *Signs of life: The growth of biotechnology centers in the U.S.* Washington, DC: Brookings Institution.

Dasgupta, Partha, and Paul A. David. 1994. Towards a new economics of science. *Research Policy* 23 (5): 487–521.

Duggan, Mark G., and William N. Evans. 2008. Estimating the impact of medical innovation: A case study of HIV antiretroviral treatments. *Forum for Health Economics and Policy* 11 (2): 1102. http://www.bepress.com/fhep/11/2/1.

Ellison, Sara Ellison, Iain M. Cockburn, Zvi Griliches, and Jerry Hausman. 1997. Characteristics of demand for pharmaceutical products: An examination of four cephalosporins. *RAND Journal of Economics* 28 (3): 426–46.

Feldman, Maryann. 2003. The locational dynamics of the U.S. biotech industry: Knowledge externalities and the anchor hypothesis. *Industry and Innovation* 10 (3): 311–28.

Finkelstein, Amy. 2004. Static and dynamic effects of health policy: Evidence from the vaccine industry. *Quarterly Journal of Economics* 119 (2): 527–64.

Freeman, Richard, and John van Reenen. 2009. What if Congress doubled R&D spending on the physical sciences? Centre for Economic Performance Discussion Paper no. 931. London: Centre for Economic Performance.

Gambardella, Alfonso. 1995. *Science and innovation: The U.S. pharmaceutical industry during the 1980s.* Cambridge, UK: Cambridge University Press.

Gans, Joshua, and Scott Stern. 2002. When does start-up innovation spur the gale of creative destruction? *RAND Journal of Economics* 33 (4): 571–86.

———. 2003. The product market and the market for "Ideas": Commercialization strategies for technology entrepreneurs. *Research Policy* 32 (2): 333–50.

Garthwaite, Craig L. 2009. Empirical essays in health economics. PhD diss., University of Maryland.

Grabowski, Henry, and John Vernon. 1996. Longer patents for increased generic competition: The Waxman-Hatch Act after one decade. *PharmacoEconomics* 10 (S2): 110–23.

Henderson, Rebecca, and Iain M. Cockburn. 1994. Measuring competence? Exploring firm effects in pharmaceutical research. *Strategic Management Journal* 15 (S1): 63–84.

Hermans, Raine, Alicia Löffler, and Scott Stern. 2008. Biotechnology. In *Innovation in global industries: U.S. firms competing in a new world,* ed. Jeffrey T. Macher and David C. Mowery. Washington, DC: National Academies Press.

Hsu, David H., and Kwanghui Lim. 2007. The antecedents and innovation consequences of organizational knowledge brokering capability. Melbourne Business School Working Paper. Carlton, Victoria, Australia: Melbourne Business School.

Jaffe, Adam B., and Josh Lerner. 2004. *Innovation and its discontents: How our broken patent system is endangering innovation and progress, and what to do about it.* Princeton, NJ: Princeton University Press.

Johnson, J. S. 1983. Human insulin from recombinant DNA technology. *Science* 219 (4585): 632–37.

Jones, Benjamin F. 2008. The burden of knowledge and the "death of the Renaissance man": Is innovation getting harder? *Review of Economic Studies* 76 (1): 283–317.

Kenney, Martin. 1986. *Biotechnology: The university-industrial complex.* New Haven, CT: Yale University Press.

Kyle, Margaret K. 2007. Pharmaceutical price controls and entry strategies. *Review of Economics and Statistics* 89 (1): 88–99.

Lerner, Josh. 1995. Patenting in the shadow of competitors. *Journal of Law and Economics* 38 (2): 463–96.

Lichtenberg, Frank R. 1995. The output contributions of computer equipment and personnel: A firm-level analysis. *Economics of Innovation and New Technology* 3 (3–4): 201–17.

———. 1998. Pharmaceutical innovation, mortality reduction, and economic growth. NBER Working Paper no. 6569. Cambridge, MA: National Bureau of Economic Research.

———. 2001. The effect of new drugs on mortality from rare diseases and HIV. NBER Working Paper no. 8677. Cambridge, MA: National Bureau of Economic Research.

———. 2005. The impact of new drug launches on longevity: Evidence from longitudinal, disease-level data from 52 countries, 1982–2001. *International Journal of Health Care Finance and Economics* 5 (1): 47–73.

———. 2008. Have newer cardiovascular drugs reduced hospitalization? Evidence from longitudinal country-level data on 20 OECD countries, 1995–2003. NBER Working Paper no. 14008. Cambridge, MA: National Bureau of Economic Research.

Lichtenberg, Frank R., and Tomas J. Philipson. 2002. The dual effects of intellectual property regulations: Within- and between-patent competition in the U.S. pharmaceuticals industry. *Journal of Law and Economics* 45 (S2): 643–72.

Lundvall, Bengt-Åke. 1992. *National systems of innovation: Towards a theory of innovation and interactive learning.* New York: St. Martin's.

Maxwell, Robert A., and Shohreh B. Eckhardt. 1990. *Drug discovery: A casebook and analysis.* Clifton, NJ: The Humana Press.

McKelvey, Maureen D. 1996. Discontinuities in genetic engineering for pharmaceuticals? Firms jumps and lock-in in systems of innovation. *Technology Analysis & Strategic Management* 8 (2): 107–16.

Merton, Robert K. 1973. *The sociology of science: Theoretical and empirical investigations,* ed. Norman W. Storer. Chicago: University of Chicago Press.

Mowery, David. 2004. *Ivory tower and industrial innovation: University-industry technology transfer before and after the Bayh-Dole Act in the United States.* Stanford, CA: Stanford Business Books.

Mowery, David C., and Richard R. Nelson. 1999. *Sources of industrial leadership: Studies of seven industries.* New York: Cambridge University Press.

Mowery, David C., and Nathan Rosenberg. 1998. *Paths of innovation: Technological change in 20th century America.* Cambridge, UK: Cambridge University Press.

Murray, Fiona. 2009. The OncoMouse that roared: Hybrid exchange strategies as a source of productive tension at the boundary of overlapping institutions. *American Journal of Sociology,* forthcoming.

Murray, Fiona, and Scott Stern. 2007. Do formal intellectual property rights hinder the free flow of scientific knowledge? An empirical test of the anti-commons hypothesis. *Journal of Economic Behavior and Organization* 63 (4): 648–87.

National Science Board. 1993. *Science and engineering indicators 1993.* Arlington, VA: National Science Foundation.

———. 2006. *Science and engineering indicators 2006.* Arlington, VA: National Science Foundation.

————. 2008. *Science and engineering indicators 2008.* Arlington, VA: National Science Foundation.

Nelson, Richard R. 1993. *National innovation systems: A comparative analysis.* New York: Oxford University Press.

Nelson, Richard R., and Nathan Rosenberg. 1994. American universities and technical advance. *Research Policy* 23 (3): 323–48.

Nelson, Richard R., and Gavin Wright. 1992. The rise and fall of American technological leadership: The postwar era in historical perspective. *Journal of Economic Literature* 30 (4): 1931–64.

Powell, Walter W., Douglas R. White, Kenneth W. Koput, and Jason Owen-Smith. 2005. Network dynamics and field evolution: The growth of interorganizational collaboration in the life sciences. *American Journal of Sociology* 110 (4): 1132–1205.

Rosenberg, Nathan. 2008. Some critical episodes in the progress of medical innovation: An Anglo-American perspective. SIEPR Discussion Paper no. 08-08. Stanford, CA: Stanford Institute for Economic Policy Research.

Scherer, F. M. 2001. The link between gross profitability and pharmaceutical R&D spending. *Health Affairs* 20 (5): 216–22.

Schwartzman, David. 1976. *Innovation in the pharmaceutical industry.* Baltimore, MD: Johns Hopkins University Press.

Stern, Scott. 1995. Incentives and focus in university and industrial research: The case of synthetic insulin. In *The university-industry interface and medical innovation,* ed. A. Gelijns and N. Rosenberg, 157–87. Washington, DC: National Academies Press.

————. 1996. Market definition and the returns to innovation: Substitution patterns in pharmaceutical markets. MIT, Sloan School of Management. Unpublished Manuscript.

————. 2004. Do scientists pay to be scientists? *Management Science* 50 (6): 835–53.

Stokes, Donald E. 1997. *Pasteur's Quadrant: Basic science and technological innovation.* Washington, DC: Brookings Institution.

Thomas, Lacy G. 1990. Regulation and firm size: FDA impacts on innovation. *RAND Journal of Economics* 21 (4): 497–517.

————. 2004. Are we all global now? Local versus foreign sources of corporate competence: The case of the Japanese pharmaceutical industry. *Strategic Management Journal* 25:865–86.

Utterback, James M. 1994. *Mastering the dynamics of innovation: How companies can seize opportunities in the face of technological change.* Boston: Harvard Business School Press.

Walsh, John P., Ashish Arora, and Wesley M. Cohen. 2003. Effects of research tool patents and licensing on biomedical innovation. In *Patents in the knowledge-based economy,* ed. Wesley M. Cohen and Stephen A. Merrill, 285–340. Washington, DC: National Academies Press.

Wuchty, Stefan, Benjamin F. Jones, and Brian Uzzi. 2007. The increasing dominance of teams in production of knowledge. *Science* 316 (5827): 1036–39.

Xu, Jiaquan, Kenneth D. Kochanek, and Betzaida Tejada-Vera. 2009. Deaths: Preliminary data for 2007. *National Vital Statistics Reports* 58 (1): 1–8. Washington, DC: National Center for Health Statistics, U.S. Department of Health and Human Services.

Zucker, Lynne G., Michael R. Darby, and Marilynn B. Brewer. 1998. Intellectual human capital and the birth of U.S. biotechnology enterprises. *American Economic Review* 88 (1): 290–306.

5

Federal Policy and the Development of Semiconductors, Computer Hardware, and Computer Software
A Policy Model for Climate Change R&D?

David C. Mowery

Advances in electronics technology in the postwar U.S. economy have created three new industries—electronic computers, computer software, and semiconductor components. These three industries also combined to give birth to the Internet, a "general purpose technology" that spans these and other industrial sectors. Electronics-based innovations supported the growth of new firms in these industries and revolutionized the operations of more mature industries, such as telecommunications, banking, and airline and railway transportation. Federal policy, especially federal research and development (R&D) investment, played a central role in the development of all of these industries.

The military applications of semiconductors and computers meant that defense-related R&D funding and procurement were important to their early development. The "R&D infrastructure" created in U.S. universities by defense-related and other federal R&D expenditures contributed to technical developments in semiconductors, computer hardware, and computer software, in addition to training a large cadre of scientists and engineers. The Internet itself emerged from federal programs that developed a national net-

David C. Mowery holds the William A. and Betty H. Hasler Chair in New Enterprise Development at the Haas School of Business, University of California, Berkeley, and is a research associate of the National Bureau of Economic Research.

This paper draws on presentations at the conference on "The Government's Role in Technology Innovation: Development of Insights for Energy Technologies," Washington, D.C., February 11, 2002, and the National Bureau of Economic Research (NBER) meetings on "Federal R&D Programs and Global Warming," November 1, 2008 and April 3, 2009. I have benefited from comments by Lewis Branscomb, Iain Cockburn, Kira Fabrizio, Rebecca Henderson, David Popp, Scott Stern, Richard Nelson, and the late Vicki Norberg-Bohm and Kenneth Sokoloff. Portions of this paper draw on Mowery and Rosenberg (1999) and Fabrizio and Mowery (2007). Research for this paper was supported in part by funding from the U.S. National Science Foundation (Cooperative Agreement #0531184).

work linking the far-flung components of the academic and industrial R&D infrastructure that had been created with federal funds. But much more than federal R&D and procurement programs were essential to the development of these technologies. Federal policies in intellectual property rights and antitrust also influenced their development, commercialization, and widespread commercial adoption of products based on them. These policies also contributed to the development of an "information technology" industry that included a large number of specialized producers of semiconductor components, computer systems, and software, in contrast to those of the European or Japanese electronics sectors, which were dominated by large, vertically integrated producers of components and systems.

Indeed, one of the most salient conclusions from the historical review presented in the following is the influence of public policies in other spheres on innovation and especially technology adoption. Although R&D programs have been valuable sources of knowledge and technological options, R&D spending alone is rarely sufficient to promote the rapid adoption of new technologies. Widespread adoption was essential to the realization of the economic benefits of innovation in electronics, and this is likely to also be true in the case of technological solutions to global warming.

Paradoxically, one important consequence of federal R&D programs and other policies in information technology (IT) was the development of a relatively weak intellectual property rights environment and in some cases, substantial interfirm technology diffusion.[1] Federal funding for procurement of the products of these new industries also encouraged the entry of new firms and interfirm technology diffusion. In addition, federal procurement supported the rapid attainment by supplier firms of relatively large production volumes, enabling faster rates of improvement in product quality and cost than otherwise would have been realized.

At least some of the catalytic effects of federal support for innovation in IT were enhanced by their "general purpose" characteristics, the rapid improvement in their price-performance ratios, and the tendency for these reductions in the price-performance ratio to accelerate adoption in a widening array of applications. In all of these technologies, the direct influence of federal R&D and procurement policies was strongest in the early years of their development, when federal expenditures on R&D or procurement accounted for the majority of such funding. Whether and how the "lessons" of these federal programs, the influence of which on innovation and industry development appears to have been greatest in the early years of technological development when the defense industry was the primary customer, can be

---

1. "Information technology" is commonly used as a summary term for a broad range of technologies including semiconductors, computer hardware and software, and telecommunications and other networking technologies. This chapter focuses on the history of the first three: Shane Greenstein (chapter 6 in this volume) explores the evolution of the Internet, perhaps the most interesting of the network technologies.

applied to the far more diverse and (in many sectors) more mature technologies relevant to climate change is an open question.

The semiconductor, computer hardware, and computer software industries now encompass many markets and applications beyond national defense, which accounts for a much smaller share of demand in all of these industries. Indeed, the technological "spillovers" that once flowed from defense-related technologies to civil applications now frequently move in the opposite direction, and the ability of Department of Defense policymakers to influence the direction of technological change has diminished considerably. Nonetheless, the substantial role of federal programs supporting innovation in the earliest stages of development of many of these industries means that the influence of these programs on intellectual property policies, interfirm technology flows, entry, and overall industrial structure remains significant today.

The electronics revolution that spawned the semiconductor and computer industries can be traced to two key innovations—the transistor and the computer. Both appeared in the late 1940s, and the exploitation of both was spurred by rapidly expanding defense spending in the early years of the Cold War, especially after the outbreak of the Korean War in 1950. The creation of these innovations also relied on domestic U.S. science and invention to a greater extent than many of the critical innovations of the pre-1940 era. The following sections briefly survey the development of the U.S. semiconductor, computer hardware, and computer software industries, highlighting the role of federal R&D and related policies in these developments.

5.1 Semiconductors

The transistor was invented at Bell Telephone Laboratories (the research arm of AT&T) in late 1947 and marked one of the first payoffs to an ambitious program of basic research in solid-state physics that Bell Labs director Mervin Kelly had launched in the 1930s. Facing increasing demands for long-distance telephone service, AT&T sought a substitute for the repeaters and relays that would otherwise have to be employed in huge numbers, greatly increasing the complexity of network maintenance and reducing reliability. Kelly felt that basic research in the emergent field of solid-state might yield suitable technologies for this purpose (Braun and MacDonald 1982).

The postwar Bell Labs R&D effort in solid-state physics, as well as others in U.S. universities (notably, the group at Purdue University headed by Karl Lark-Horovitz) built on extensive wartime R&D in electronics and radar that had explored the properties of semiconducting materials such as germanium. Much of this R&D was managed by the Massachusetts Institute of Technology (MIT) Radiation Laboratory, which supported the Lark-Horovitz research team. After 1945, the U.S. Signal Corps continued to support Lark-Horovitz and his colleagues, who were pursuing research in semiconducting amplifiers and were seen by the Bell Labs research team as a significant threat in the race to develop the first semiconductor-based amplifier.[2]

Bell Labs' commercial exploitation of its discovery was constrained in various ways by the antitrust suit against AT&T filed in 1949 by the U.S. Department of Justice. Faced with this threat to its existence, AT&T was reluctant to develop an entirely new line of business in the commercial sale of transistor products, and it may have wished to avoid any practice that would draw attention to its market power, such as charging high prices for transistor components or patent licenses. In addition, the military services that had begun to support Bell Labs' transistor research also encouraged the dissemination of transistor technology. In September 1951, a symposium was held at Bell Labs and attended by 139 industrial representatives, 121 military personnel, and 41 university scientists. The proceedings of this symposium were widely distributed to Bell licensees and others, aided by financial assistance from the U.S. military, which distributed 3,000 copies at public expense (Misa 1985). A 1952 symposium for attendees who had paid a $25,000 licensing fee focused on transistor production techniques for the point-contact transistor and produced two thick volumes on semiconductor technology, known within Bell Labs as "the Bible" (Misa 1985).[3]

The federal antitrust suit was settled through a consent decree in 1956, and AT&T restricted its commercial activities to telecommunications service and equipment. The 1956 consent decree also led AT&T, holder of a dominant patent position in semiconductor technology, to license its semiconductor patents at nominal rates to all comers, seeking cross-licenses in exchange for access to its patents. As a result, virtually every important technological development in the industry was accessible to AT&T, and all of the patents in the industry were linked through cross-licenses with AT&T.

The transistor had important potential military applications in military

2. According to Riordan and Hoddeson (1997, 162), following a private demonstration of the transistor by the Bell Labs research team for senior military researchers in June 1948:

> Shockley buttonholed Harold Zahl of the Army Signal Corps, which had been funding most of Purdue's research on germanium. "Tell me one thing, Harold," he asked him impetuously. "Have Lark-Horovitz and his people at Purdue already discovered this effect, and perchance has the military put a TOP SECRET wrap on it?" A great expression of relief came over Shockley's face when Zahl told him that the Purdue physicists had not, although they were probably only six months away. Recalled Zahl, "Bill was happy, for to him six months was infinity!"

As this anecdote suggests, senior Bell Labs management, including Kelly, were concerned that in spite of the lack of direct funding from the Department of Defense for their work on transistors, the invention might be classified and its application to civilian uses (and markets) restricted. One reason for the private "demonstration" of the transistor was to ascertain whether the military would insist on classification. In the event, no such demands were made.

3. Holbrook et al. (2000) point out that the 1952 symposium focused on production technologies for the point-contact transistor, which was soon to be superseded by the junction transistor that Shockley had invented subsequently to the team's invention of the point-contact device. As a result, much of the production know-how disseminated at the second symposium proved to be obsolete.

electronics and computer systems. Moreover, the invention appeared just as the nascent Cold War was warming up considerably. By 1950, the Korean War and the Soviet explosion of an atomic bomb had triggered rapid growth in U.S. defense spending as part of a long-term shift to a much larger defense establishment that focused on strategic nuclear weapons and measures to defend against strategic airborne threats.[4] Considerable process R&D and "trial and error" experimentation were needed to support volume production of transistors, and military spending on industrial R&D focused on the development of production technologies. By 1953, the U.S. Department of Defense was funding pilot transistor production lines operated by AT&T, General Electric, Raytheon, Sylvania, and RCA (Tilton 1971). As figure 5.1 shows, R&D spending per se initially accounted for a relatively small share of federal technology-development contracts, reflecting the importance of production engineering and the focus of military policymakers on expanding production of even the relatively primitive early transistors. Defense-related expenditures also supported the construction of large-scale production facilities, including a large-scale Western Electric transistor plant in Pennsylvania. In combination with the political environment of near-wartime mobilization, such large-scale public funding commitments to production as well as R&D activities may have assured industrial firms of the depth and credibility of the federal commitment to this technology, encouraging complementary private investments in transistor development and production.

The R&D share of federal spending through these contracts rapidly grew, however, and by 1959 they accounted for more than 80 percent of federal spending in semiconductor-related technology development within these firms. But overall, defense-related federal R&D spending appears to have been focused on more applied activities. In this respect, the profile of defense-related R&D spending in electronics resembled the overall composition of the national defense-related R&D budget, which historically has been dominated by development activities.

According to Tilton (1971), federally supported R&D accounted for nearly 25 percent of total semiconductor-industry R&D spending in the late 1950s. Interestingly, the bulk of this federal R&D spending during the 1950s was allocated to established producers of electronic components. Indeed, Tilton (1971) shows that "new firms," including Texas Instruments, Shockley Laboratories, Transitron, and Fairchild, received only 22 percent of federal R&D contracts in 1959. These new firms nonetheless accounted for 63 percent of semiconductor sales in that year. Defense procurement contracts, which frequently were awarded to new firms, proved to be at least as important as public funding of industry R&D in shaping this nascent industry.

---

4. This shift in U.S. defense policy and spending also benefited the computer hardware industry, as I note later is this chapter.

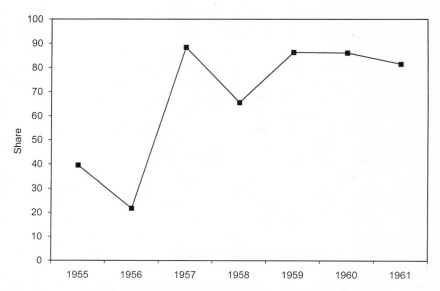

Fig. 5.1   R&D share of federal semiconductor development contracts to firms, 1955–1961
Source: Tilton (1971).

With few exceptions, these "new" firms were founded by employees of established electronics firms (such as AT&T or, eventually, Shockley Semiconductor) or (as in the case of Texas Instruments) diversified into the nascent semiconductor industry from other lines of business (see Klepper 2009). University-based researchers, with the exception of the Purdue research group and a few others, played a relatively modest role in these early innovations. Conversely, Shockley Semiconductor's fabrication facility near Stanford University played an important role as a site for research by Stanford faculty on fabrication techniques for semiconductors. For example, James Gibbons, who later became Stanford's dean of engineering, worked at Shockley Semiconductor while a junior faculty member at Stanford. Gibbons was sent there by the university's provost and leader of the Solid-State Laboratory to "learn the techniques required for the fabrication of silicon devices from Shockley and then transfer these techniques back to the university" (Lécuyer 2005, 138).

The first commercially successful transistor was produced by Texas Instruments (rather than AT&T) in 1954. Texas Instruments' silicon junction transistor was quickly adopted by the U.S. military for use in radar and missile applications. The next major advance in semiconductor electronics was the integrated circuit, which combined a number of transistors on a single silicon chip, in 1958. The integrated circuit was invented by Jack Kilby of Texas Instruments and drew on that company's innovations in diffusion

and oxide masking technologies that were first developed for the manufacture of silicon junction transistors. The development of the integrated
circuit made possible the interconnection of large numbers of transistors on
a single device, and its commercial introduction in 1961 spurred tremendous
growth in the industry's sales.

Kilby's search for the integrated circuit was motivated by his employer's
interest in producing a device that could expand the military (and, eventually,
the commercial) market for semiconductors. Little of Kilby's pathbreaking
R&D was supported by the U.S. military, but the military was a large-scale
purchaser of integrated circuits once they became available.[5] Indeed, the
prospect of large procurement contracts appears to have operated similarly
to a prize, leading Texas Instruments to invest its own funds in the development of a product that met military requirements. Figure 5.2 highlights
the significant share of integrated circuit (IC) shipments accounted for by
government purchases in the early years of the industry's history, as well
as the decline in this share as commercial markets for the IC grew during
the 1960s. A longer time series for the government share of semiconductor
shipments (figure 5.3) similarly shows the importance of government procurement in the early years of the broader semiconductor industry, as well
as the shrinking share of demand represented by federal procurement after
the 1960s. By the 1990s, military demand accounted for less than 10 percent
of integrated circuit sales (figure 5.4).

One result of government involvement in the early postwar semiconductor industry as both a funder of R&D and a purchaser of its products was
the emergence of a new structure for the innovation and technology commercialization processes. This new structure contrasted with that of pre-
1940 technology-intensive U.S. industries, such as chemicals or electrical
machinery, or the semiconductor industries of such nations as Germany
or Japan. In a virtual reversal of the prewar situation, large U.S. firms were
much more significant as R&D performers than in producing and selling
new semiconductor devices. The entrant firms' role in the introduction of

---

5. Malerba's (1985) discussion of the development of the Western European and U.S.
semiconductor industries emphasizes the importance of the large scale of military R&D and
procurement programs in the United States, as well as the focus of defense-related R&D on
industry performers:

> The size of American [R&D] support was much greater than that of either the British or the
> European case generally, but particularly during the 1950s. Second, the timing of policies
> was different: while the United States was pushing the missile and space programs in the
> second half of the 1950s/early 1960s, Britain was gradually retreating from such programs.
> Third, American policies were more flexible and more responsive than British policies. Fi-
> nally, research contracts in the United States focused more on development than on re-
> search, while in Britain, as well as in the rest of Europe, such contracts focused more on
> research and proportionately more funds were channeled into government and university
> laboratories. These last two factors meant that most R&D projects in Britain, as well as in
> Europe, were not connected with the commercial application of the results of R&D. (Mal-
> erba 1985, 82)

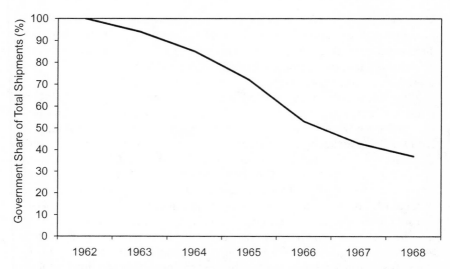

Fig. 5.2   Government purchases of integrated circuits as a share of total shipments
Source: Levin (1982, table 2.17, 63).

Original Sources: Tilton (1971), 91. Total shipments data originally drawn from Electronic Industries Association, *Electronic Industries Yearbook, 1969* (Washington, DC: Electronic Industries Association, 1969). Government share calculated by Tilton from data in Business and Defense Services Administration (BDSA), "Consolidated Tabulation: Shipment of Selected Electronic Components."

Notes: The following notes are from Levin (1982, table 2.17, 63).

a. Includes circuits produced for Department of Defense, Atomic Energy Commission, Central Intelligence Agency, Federal Aviation Agency, and National Aeronautics and Space Administration.

b. Estimated by Tilton (1971).

new products, reflected in their often-dominant share of markets in new semiconductor devices, significantly outstripped that of larger firms. Moreover, the role of new firms grew in importance with the development of the integrated circuit.

Although the military market for integrated circuits was soon overtaken by commercial demand, military demand spurred the early industry growth and price reductions that created a large commercial market for integrated circuits. The large volume of integrated circuits produced for the military market allowed firms to move rapidly down learning curves, reducing component costs sufficiently to create a strong commercial demand.[6] According to Tilton (1971), a doubling of cumulative output produced a 20 to 30 percent drop in the costs of production for these early semiconductor devices. During 1962 to 1978, total shipments of ICs to governments fell from 100

6. From the beginning, Texas Instruments and other firms were aware of the commercial potential of the integrated circuit. As one of its first demonstration projects, Texas Instruments constructed a computer to demonstrate the reductions in component count and size that were possible with integrated circuits.

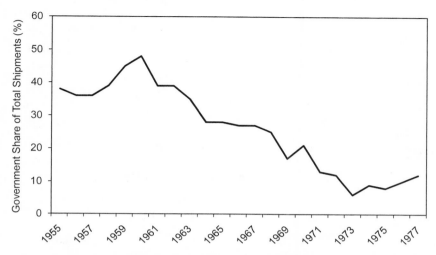

Fig. 5.3   Government purchases of semiconductor devices as share of total shipments
Source: Levin (1982, table 2.16, 63).

Original Sources: 1952 to 1959 data from U.S. Department of Commerce, Business and Defense Services Administration, *Electronic Component: Production and Related Data, 1952–1959* (Washington, DC: U.S. Department of Commerce, 1960). 1960 to 1968 data from BDSA, "Consolidated Tabulation: Shipments of Selected Electronic Components," Mimeograph (Washington, DC: BDSA, annually). 1969 to 1977 data from U.S. Department of Commerce, Bureau of the Census, Current Industrial Reports, Series MA-175, "Shipments of Defense-Oriented Industries" (Washington, DC: U.S. Department of Commerce, annually).

Notes: The following notes are from Levin (1982, table 2.16, 63). Includes devices produced for Department of Defense, Atomic Energy Commission, Central Intelligence Agency, Federal Aviation Agency, and National Aeronautics and Space Administration equipment.

percent to 10 percent, while the share of shipments to industrial and commercial users rose from 0 to 90 percent (table 5.1).

Military procurement policies also influenced industry structure. In contrast to Western European defense ministries, which directed the bulk of their R&D funding and procurement funding to established defense suppliers (Flamm 1988, 134), the U.S. military was willing to award substantial procurement contracts to firms, such as Texas Instruments, that had recently entered the semiconductor industry but had little or no history of supplying the military. The U.S. military's willingness to purchase from untried suppliers was accompanied by conditions that mandated substantial technology transfer among U.S. semiconductor firms. To reduce the risk that a system designed around a particular integrated circuit would be delayed by production problems or by the exit of a supplier, the military required its suppliers to develop a "second source" for the product—that is, a domestic producer that could manufacture an electronically and functionally identical product. To comply with second-source requirements, firms exchanged designs and shared sufficient process knowledge to ensure that the component produced by a second source was identical to the original product.

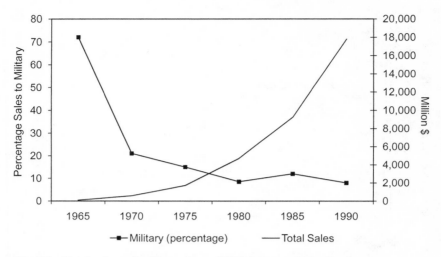

Fig. 5.4   Total sales and military share of U.S. integrated circuit sales

Source: Alic et al. (1992, table 8-1, 260).

Original Sources: 1965, 1970: Normal J. Asher and Leland D. Strom, "The Role of the Department of Defense in the Development of Integrated Circuits," IDA Paper no. P-1271 (Arlington, VA: Institute for Defense Analyses, May 1977), 73; 1975: Estimated, based on *A Report on the U.S. Semiconductor Industry* (Washington, DC: Department of Commerce, September 1979), 39, 44; 1980: *As Assessment of the Impact of the Department of Defense Very High Speed Integrated Circuit Program,* National Materials Advisory Board Report no. NMAB-382 (Washington, DC: National Research Council, January 1982), 64; 1985: *Report of the Defense Science Board on Use of Commercial Components in Military Equipment* (Washington, DC: Office of the Under Secretary of Defense for Acquisition, June 1989), A-14; 1990: Estimated, based on figures from Dataquest and the Semiconductor Industry Association.

Table 5.1   **End-use shares of total U.S. sales of integrated circuits and total market value, 1962–1978**

Market	1962	1965	1969	1974	1978
Government (%)	100	55	36	20	10
Computer (%)	0	35	44	36	37.5
Industrial (%)	0	9	16	30	37.5
Consumer (%)	0	1	4	15	15
Total U.S. domestic shipments ($ millions)	4	79	413	1,204	2,080

Sources: Langlois & Steinmueller (1999, 37, table 2.7); Borrus, Millstein, and Zysman (1983, 159).

By facilitating entry and supporting high levels of knowledge diffusion among firms, public policy (e.g., the 1956 AT&T consent decree, the Department of Defense "second source" policy) and other influences increased the diversity and number of technological alternatives explored by individuals and firms within the U.S. semiconductor industry during a period of significant uncertainty about the direction of future development of this technology. Extensive entry and rapid interfirm technology diffusion also fed intense competition among U.S. firms. The competitive industry structure and conduct enforced a rigorous "selection environment," ruthlessly weeding out less-effective firms and technical solutions. For a nation that was pioneering in the semiconductor industry, this combination of technological diversity and strong selection pressures proved to be highly effective.

As nondefense demand for semiconductor components grew and came to dominate industry demand, defense-to-civilian technology "spillovers" declined in significance and actually reversed in direction. By the 1970s, a combination of rapid product innovation in civilian technologies and longer delays in military procurement programs meant that military-specification semiconductor components often lagged behind their commercial counterparts in technical performance, although these "milspec" components could operate in more "hostile" environments of high temperatures or vibration. Nonetheless, concern among U.S. defense policymakers over this "technology gap" grew and resulted in the creation of the Department of Defense's Very High Speed Integrated Circuit (VHSIC) program in 1980. Federally funded VHSIC projects linked merchant semiconductor firms largely devoted to commercial production with semiconductor equipment manufacturers and defense-systems producers in development projects intended to produce advanced, high-speed, "milspec" components.

Originally planned for a six-year period and budgeted at slightly more than $200 million, the VHSIC program lasted for ten years and cost nearly $900 million. Nonetheless, the program failed to meet its objectives, demonstrating the limited influence of the federal government within a U.S. semiconductor market that by the 1980s was dominated by commercial applications and products. In the wake of the VHSIC program's mixed results, some scholars have recommended that defense policymakers seek to change procurement policies to enable more rapid incorporation of commercial innovations (Alic et al. 1992).

Another federally funded R&D initiative in the semiconductor industry that highlights the changing relationship between civilian and defense-related innovation was the Semiconductor Manufacturing Technology Consortium (SEMATECH). Founded in 1987, SEMATECH supported collaborative R&D among leading U.S. semiconductor firms on manufacturing processes in an effort to improve manufacturing performance in the face of intense competition from Japanese producers.[7] The Semiconductor Manufacturing Technology Consortium initially received one-half of its $200 million annual operating budget from the federal government, based on arguments by the Defense Science Board and other experts (U.S. Department of Defense 1987; Alic et al. 1992) claiming that the U.S. civilian semi-

---

7. The founding members of SEMATECH were Advanced Micro Devices, AT&T, Digital Equipment Corporation, Harris Corporation, Hewlett-Packard Company, Intel Corporation, IBM, LSI Logic, Micron Technology, Motorola, National Semiconductor, NCR, Rockwell International, and Texas Instruments. Micron and LSI Logic withdrew in 1991, and Harris

conductor "industrial base" was essential to the nation's defense establishment.[8] Based on the assertion that defense-related procurement alone could no longer sustain a viable U.S. semiconductor industry, defense funds were used to support R&D on manufacturing technologies that were relevant to civilian products, many (but not all) of which had applications in defense systems as well.

Organized in considerable haste, the consortium's original R&D agenda proved to be unsustainable. The original vision of sharing sensitive manufacturing know-how among firms that were direct competitors in many markets was never realized, and SEMATECH gradually shifted its focus to supporting collaboration between semiconductor manufacturers and producers of semiconductor processing equipment. As SEMATECH Chief Executive Officer William J. Spencer remarked in 1992, "We can't develop specific products or processes. That's the job of the member companies. SEMATECH can enable members to cooperate or compete as they see fit" (Burrows 1992, 58). This approach proved to be more viable although the consortium's support did not prevent the exit of some important U.S. equipment producers, and SEMATECH member firms were not consistently willing to follow their investments in R&D with purchases of commercial models of the equipment developed under the consortium's sponsorship. By 1997, SEMATECH had invited non-U.S. semiconductor manufacturing firms to become members (although Japanese firms did not join a subsidiary of the consortium until 2004), and federal funding had ended.

The Semiconductor Manufacturing Technology Consortium's effects on the U.S. semiconductor industry remain controversial (see Macher, Mowery, and Hodges 1999). Although the market shares of both U.S. manufacturers and equipment suppliers began to improve by the early 1990s, much of this improvement reflected the decline in the fortunes of Japanese manufacturers that occurred during this period, along with the entry and expansion of South Korean and Taiwanese semiconductor manufacturers, who were significant purchasers of U.S. equipment firms' products. Nevertheless, leading U.S. (and non-U.S.) manufacturers and equipment suppliers remain active in the consortium, and their willingness to maintain their investments in its R&D activities suggests that they find its "vertical" R&D strategy to be valuable. During and since its support from federal sources, SEMATECH's R&D program has not focused on basic research. Instead, the consortium has emphasized the collective development by manufacturers and suppliers of technology "roadmaps" over a five-year future to guide R&D investments and product development, as well as facilitating agreement on technical standards and performance goals for equipment.

Corporation withdrew in 1992. At the time of its formation, the fourteen original members of SEMATECH accounted for more than 80 percent of U.S. semiconductor production capacity.

8. See Grindley, Mowery, and Silverman (1994) for a more extensive discussion of SEMATECH.

5.2   Computer Hardware

The development of the U.S. computer industry also benefited from Cold War military spending, but in other respects, the origins and early years of this industry differed from semiconductors. One marked difference was the role of U.S. universities in the industry. Although they were peripheral actors in the early development of semiconductor technology, U.S. universities were important sites for the early development, as well as the research activities that produced the earliest U.S. electronic computers. In addition, federal spending during the late 1950s and 1960s from military and nonmilitary sources provided an important basic research and educational infrastructure for the development of this new industry and the broader IT sector.

During World War II, the American military sponsored a number of projects to develop high-speed calculators to solve special military problems. The ENIAC—generally considered the first fully electronic U.S. digital computer—was funded by Army Ordnance as a device for computing firing tables for artillery. Developed by J. Presper Eckert and John W. Mauchly at the University of Pennsylvania, the ENIAC required rewiring for each new problem. In 1944, John von Neumann began advising the Eckert-Mauchly team, which was then working on the development of a new machine, the EDVAC. This collaboration developed the concept of the stored-program computer: instead of being wired for a specific problem, the EDVAC's instructions were stored in memory, facilitating their modification.

Von Neumann's abstract discussion of the concept (von Neumann [1945] 1986) circulated widely and served as the logical basis for subsequent computers. Indeed, the extensive dissemination of the EDVAC report led U.S. Army patent attorneys to rule that its basic ideas were not patentable, spurring the broad exploitation of this fundamental architectural innovation (Flamm 1988). The subsequent settlement in 1956 of a federal antitrust suit against IBM also included liberal licensing decrees, further supporting interfirm diffusion of computer technology.

The first fully operational stored-program computer in the United States was the SEAC, built by the National Bureau of Standards in 1950 (Flamm 1988). A number of other important machines were developed for or initially sold to federal agencies, including the Princeton IAS computer, built by von Neumann at the Institute for Advanced Study in 1951, MIT's Whirlwind computer, developed in 1949, and the UNIVAC, built in 1953 by Remington Rand based on the Eckert-Mauchly technology.[9] At least nineteen government-funded development projects produced electronic computers during the 1945 to 1955 period (Flamm 1988). (See table 5.2.)

From the earliest days of their support for the development of computer technology, the U.S. armed forces sought to ensure that technical informa-

---

9. Table 5.2 contains a more complete listing of these early postwar government-supported computer development projects.

Table 5.2   **Early federal government computer-development programs**

First generation of U.S. computer projects	Estimated cost of each machine ($ thousands)	Source of funding	Initial operation
ENIAC	750	Army	1945
Harvard Mark II	840	Navy	1947
Eckert-Mauchly BINAC	178	Air Force (Northrop)	1949
Harvard Mark III	1,160	Navy	1949
NBS Interim computer (SEAC)	188[a]	Air Force	1950
ERA 1101 (Atlas I)	500	Navy/NSA[b]	1950
Eckert-Mauchly UNIVAC	400–500	Army via census; Air Force	1951
MIT Whirlwind	4,000–5,000	Navy; Air Force	1951
Princeton IAS computer	650[a]	Army; Navy; RCA; AEC	1951
University of California CALDIC	95[a]	Navy	1951
Harvard Mark IV	n.a.	Air Force	1951
EDVAC	467	Army	1952
Raytheon Hurricane (RAYDAC)	460[a]	Navy	1952
ORDVAC	600	Army	1952
NBS/UCLA Zephyr computer (SWAC)	400	Navy; Air Force	1952
ERA Logistics computer	350–650	Navy	1953
ERA 1102 (3 built)	1,400[c]	Air Force	1953
ERA 1103 (Atlas II, 20 built)	895	Navy/NSA[b]	1953
IBM Naval Ordnance Research Computer (NORC)	2,500	Navy	1955

Source: Flamm (1988, 76)

Original Sources: Herman H. Goldstine, *The Computer from Pascal to von Neumann* (Princeton, NJ: Princeton University Press, 1972), 242–45, 316–18, 326, 328; Arthur D. Little, Inc., with the White, Weld & Co. research department, *The Electronic Data Processing Industry: Present Equipment, Technological Trends, Potential Markets* (New York: White, Weld & Co., 1956), 82; Martin H. Weik, "A Third Survey of Domestic Electronic Digital Computing Systems," Report no. 115 (Aberdeen Proving Ground, MD: Ballistic Research Laboratories, 1961), 213, 236, 282, 393, 567, 635, 639, 676–77, 732, 848, 900, 1016, 1081–83; Martin H. Weik, "A Fourth Survey of Domestic Electronic Digital Computing Systems," Report no. 1227 (Aberdeen Proving Ground, MD: Ballistic Research Laboratories, 1964), 373; Nancy Stern, *From ENIAC to UNIVAC: An Appraisal of the Eckert-Mauchly Computers* (Bedford, MA: Digital Press, 1981), 37, 51, 62, 105, 113, 117, 122–23, 132; Kent C. Raymond and Thomas M. Smith, *Project Whirlwind: The History of a Pioneer Computer* (Bedford, MA: Digital Press, 1980), 107, 110, 127–28, 156–58, 166; Ralph A. Niemann, *Dahlgren's Participation in the Development of Computer Technology* (Dahlgren, VA: Naval Surface Weapons Center, 1982), 4, 5, 11, 16; Samuel S. Snyder, "Influence of U.S. Cryptologic Organizations on the Digital Computer Industry," SRH 003, declassified National Security Agency report released to the National Archives, 7; Samuel S. Snyder, "Computer Advances Pioneered by Cryptologic Organizations," *Annals of the History of Computing,* vol. 2 (January 1980), 60–63; M. R. Williams, "Howard Aiken and the Harvard Computation Laboratory," *Annals of the History of Computing,* vol. 6 (April 1984), 160; ONR, *Digital Computer Newsletter,* various issues, 1949–1956; S. N. Alexander, "Introduction," in U.S. Department of Commerce, National Bureau of Standards, Computer Development (SEAC and DYSEAC) at the National Bureau of Standards, Washington, D.C., NBS circular no. 551 (Washington, DC: GPO, 1955), 3; H. D. Huskey, "The National Bureau of Standards Western Automatic Computer (SWAC)," *Annals of the History of Computing,* vol. 2 (April 1980), 111–21; John W. Carr III, "Instruction Logic of the MIDAC," in C. Gordon Bell and Allen Newell, eds., *Computer Structures: Readings and Examples* (McGraw-Hill, 1971), 209; John Varick Wells, "The Origins of the Computer Industry: A Case Study in Radical Technological Change" (PhD dissertation, Yale University, 1978), 268; and the following citations in N. Metropolis, J. Howlettm and Gian-Carlo Rota, eds., *A History of Computing in the Twentieth Century: A Collection of Essays* (Academic Press, 1980): J. C. Chu, "Computer Development at Argonne National Laboratory," 346; James E. Robertson, "The

Table 5.2   (continued)

ORDVAC and the ILLIAM," 346–47; Henry D. Huskey, "The SWAC: The National Bureau of Standards Western Automatic Computer," 421, 428, 430; M. Metropolis, "The MANIAC," 462; and Erwin Tomash, "The Start of an ERA: Engineering Research Associates, Inc., 1946–1955," 491.

Note: n.a. = not available.

[a]Estimated cost in 1950, in "Report on Electronic Digital Computers by the Consultants to the Chairmand of the Research and Development Board," June 15, 1950, appendix 4, cited by Kent C. Redmond and Thomas M. Smith (1980, 166).

[b]The National Security Agency (NSA) includes Army and Navy predecessor agencies.

[c]Cost for three machines.

tion on this innovation reach a broad industrial and academic audience. This attitude, which contrasted with that of the militaries of the United Kingdom and the Soviet Union, appears to have stemmed from the U.S. military's concern that a substantial industry and research infrastructure would be required for the development and exploitation of computer technology.[10] The technical plans for the military-sponsored IAS computer were widely circulated among U.S. government and academic research institutes, and it spawned a number of "clones," including the ILLIAC, the MANIAC, AVIDAC, ORACLE, and JOHNIAC (Flamm 1988).

By 1954, the ranks of the leading U.S. computer manufacturers were dominated by established firms in the office equipment and consumer electronics industries, including RCA; Sperry Rand (originally the typewriter producer Remington Rand, which had acquired Eckert and Mauchly's embryonic computer firm); NCR; and IBM. These firms focused on the business market as well as scientific computing, while Bendix Aviation acquired the computer operations of Northrop Aircraft and specialized in computers for scientific applications (Flamm, 1988, 82). The National Security Agency, the Atomic Energy Commission, and the Department of Defense all supported the development of advanced computer systems for specialized applications in air defense, cryptography, and nuclear weapons design.

IBM's technology development efforts benefited from the firm's experience as supplier of more than fifty large computers for the Semi-Automatic

10. Herman Goldstine (1993, 217), one of the leaders of the wartime project sponsored by the Army's Ballistics Research Laboratory at the University of Pennsylvania that resulted in the Eckert-Mauchly computer, notes:

A meeting was held in the fall of 1945 at the Ballistic Research Laboratory to consider the computing needs of that laboratory "in the light of its post-war research program." The minutes indicate a very great desire at this time on the part of the leaders there to make their work widely available. "It was accordingly proposed that as soon as the ENIAC was successfully working, its logical and operational characteristics be completely declassified and sufficient be given to the machine . . . that those who are interested . . . will be allowed to know all details."

Goldstine is quoting the "Minutes, Meeting on Computing Methods and Devices at Ballistic Research Laboratory," 15 October 1945 (note 14).

Ground Environment (SAGE) air-defense network that was developed under the supervision of MIT's Lincoln Laboratories in the 1950s, and the firm was awarded a contract by the Atomic Energy Commission in 1956 for an advanced computer (referred to as the "Stretch" project) for use by Los Alamos National Laboratories. Other U.S. computer firms, including Sperry Rand and ERA, produced advanced computers in small quantities for federal intelligence and defense agencies during the 1950s. According to Flamm (1987), federal funds accounted for 59 percent of the combined computer-related R&D spending of General Electric, IBM, Sperry Rand, AT&T, Raytheon, RCA, and Computer Control Corporation during 1949 to 1959. Even within the mature U.S. computer industry, federal funds accounted for a significant share of overall R&D activity. Flamm (1987, 243) estimates that federal funds accounted for almost 13 percent of total R&D in the "office, computing, and accounting machines" industry in 1981, increasing to slightly more than 15 percent by 1984.

Business demand for computers gradually expanded during the early 1950s to form a substantial market. The most commercially successful machine of the decade, with sales of 1,800 units, was the IBM 650 (Fisher, McKie, and Mancke 1983). Even in the case of the IBM 650, however, government procurement was still crucial: the projected sale of 50 machines to the federal government (a substantial portion of the total forecast sales of 250 machines) influenced IBM's decision to move the computer into full-scale development (Flamm 1988). Government purchase made up a substantial portion of IBM sales during the 1950s, but they became proportionally less important through the following two decades as private-sector sales grew (figure 5.5). Although the commercial operations of IBM and other early U.S. computer producers benefited from extensive federal R&D and procurement funding, IBM's increasing dominance of commercial computer sales by the late 1950s also drew on the marketing and manufacturing capabilities that had been developed through the firm's long history as a major producer of office equipment.[11]

The federal share of overall computer-industry sales had declined significantly by the end of the 1950s (by 1966, federal government installations accounted for no more than 10 percent of the total U.S. installed base of com-

11. "Like most 'new' technology, the computer had important antecedents, and existing firms had developed capabilities related to those antecedents. None had done so more thoroughly than IBM. Despite some superficial differences between computers and the earlier tabulating equipment that had formed the core of IBM's business, computers involved a mix of knowledge and capabilities that matched those existing at IBM extraordinarily well" (Usselman 1993, 5). Usselman also emphasized the skills at IBM's Endicott, New York factory that had long been the source of many of the firm's tabulating machines: "The Endicott facility also produced a series of input-output devices that helped develop the market for both large and small computers. Though these products made use of electronics, they also drew on the mechanical skills available at Endicott. Printers and disk storage devices in particular were as distinguished [sic] as much for their rapid, precise mechanical motions as for their logical design" (Usselman 1993, 12). Chandler (2001) makes a similar argument.

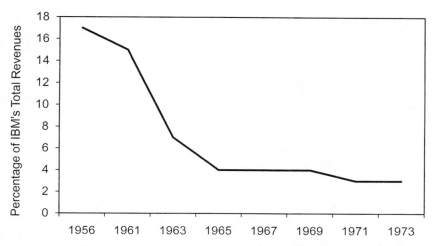

Fig. 5.5   IBM sales of special products and services to U.S. government agencies
Source: Flamm (1987, table 4-7, 108).

Original Sources: Montgomery Phister, Jr., *Data Processing Technology and Economics,* 2nd
ed. (Bedford, MA: Digital Press and Santa Monica Publishing, 1979), 310.

puters, according to Flamm [1987]), but government purchases accounted
for a larger share of high-performance computer sales. In 1972, more than
40 percent of total sales of computers in the highest-performance class
("Class 7," which includes supercomputers, along with other less-advanced
systems) went to the federal government. By 1980, this share was still slightly
more than 13 percent. Within supercomputers alone, the federal share of
overall purchases was significantly higher.

Even after the emergence of a substantial private industry dedicated to the
development and manufacture of computer hardware, federal R&D funding aided the creation of the new academic discipline of computer science.
In addition to their role as sites for applied and basic research in computer
hardware and software, U.S. universities produced engineers and scientists
active in the computer industry (see the following for further discussion). By
virtue of their relatively "open" research and operating environments that
emphasized publication, relatively high levels of turnover among research
staff, and the production of graduates who sought employment in industry,
universities served as sites for the dissemination and diffusion of innovations throughout the industry. United States universities provided important
channels for cross-fertilization and information exchange between industry
and academia and also between defense and civilian research efforts in software and in computer science generally.

The institution-building efforts of the National Science Foundation
(NSF) and the Department of Defense came to overshadow private-sector
contributions by the late 1950s. In 1963, about half of the $97 million spent

by universities on computer equipment came from the federal government, while the universities themselves paid for 34 percent, and computer makers underwrote the remaining 16 percent (Fisher, McKie, and Mancke 1983). Federal funding for computer-related research accounted for a significant portion of the total (including industry- and university-funded) R&D performed outside of industry through the 1980s (see figure 5.6). During the 1970s and 1980s, roughly 75 percent of the mathematics and computer science research performed at universities was funded by the federal government (Flamm 1987).

According to a recent report from the National Research Council's Computer Science and Telecommunications Board, federal investments in computer science research increased fivefold during the 1976 to 1995 period, from $180 million in 1976 to $960 million in 1995 (in constant 1995 dollars). Federally funded basic research in computer science, roughly 70 percent of which was performed in U.S. universities, grew from $65 million in 1976 to $265 million in 1995 (National Research Council 1999). The defense share of federal computer science research funding declined from almost 70 percent

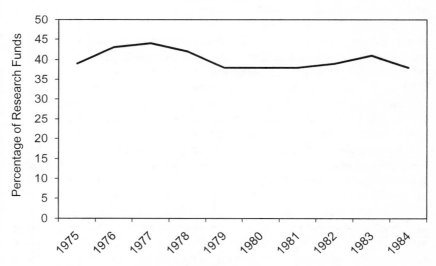

Fig. 5.6   Federal math and computer science funds as a share of all computer-related research performed in universities and nonindustrial research organizations

Source: Flamm (1987, table 4-5, 104).

Original Sources: National Science Foundation (NSF), *Research and Development in Industry, 1984* (Washington, DC: NSF, 1985), 20, 23; NSF, *Federal Funds for Research and Development: Federal Obligations for Research by Agency and Detailed Field of Science, Fiscal Years 1967–86* (Washington, DC: GPO, 1985), 5, 31; NSF, *Academic Science/Engineering: R&D Funds, Fiscal Year 1982,* NSF 84-308 (Washington, DC: GPO, 1984), 129–30; NSF data obtained though computer database; NSF, *Academic Science/Engineering: 1972–83* (Washington, DC: NSF, 1984), 43–44; and NSF, *Academic Science/Engineering: R&D Funds, Fiscal Year 1983* (Washington, DC: NSF, 1984), 16, 130–31.

in fiscal 1976 to slightly more than 40 percent by fiscal 1996 (Clement 1987, 1989; Clement and Edgar 1988), and defense funding of computer science research in universities appears to have been supplanted somewhat by the growth in funding for quasi-academic research and training organizations.

The federal government's R&D spending was supplemented by procurement spending on systems for military applications. In both the hardware and software areas, the government's needs differed from those of the commercial sector, and the magnitude of purely technological "spillovers" from military R&D and procurement to civilian applications appear to have declined somewhat as the computer industry moved into the 1960s. Just as had been the case in semiconductors, military procurement demand acted as a powerful attraction for new firms to enter the industry, and many such enterprises entered the fledgling U.S. computer industry in the late 1950s and 1960s. Antitrust policy played a role here as well—another 1956 consent decree, this time settling an antitrust suit filed by the federal government against IBM, resulted in extensive licensing of the firm's patents at low royalties. Although IBM's position in computer technologies by this date was not as dominant as that of AT&T in semiconductors, the computer firm nevertheless had an extensive patent portfolio that became much more widely available to other firms through low-cost licensing agreements as a result of the 1956 settlement of the antitrust suit. Moreover, it is likely that IBM's willingness to pursue alleged infringers of its patents was curtailed by this federal suit and its settlement.

5.3   Computer Software

By the 1980s, the development of the semiconductor and computer industries had laid the groundwork for the expansion of another "new" industry, the production of standardized computer software for commercial markets (as opposed to the commercial production of custom software or user-developed custom software).[12] The growth of the U.S. computer software industry has been marked by at least four distinct eras. During the earliest years of the first era (1945 to 1965), covering the development and early commercialization of the computer, software as it is currently known did not exist. The concept of computer software as a distinguishable component of a computer system was effectively born with the advent of the von Neumann architecture for stored-program computers. But even after the von Neumann scheme became dominant in the 1950s, software remained closely bound to hardware, and the organization producing the hardware generally developed the software as well. As computer technology developed and the market for its applications expanded after 1970, however, users, indepen-

---

12. A more detailed discussion of the U.S. and other industrial nations' software industries, on which this section draws, may be found in Mowery (1999).

dent developers, and computer service firms began to play prominent roles in software development.

The development of a U.S. software industry began only when computers began to be adopted on a large scale in commercial uses, a development spurred by the success of the IBM 650. Widespread adoption of a single computer platform contributed to the growth of "internal" software production by large users. But the primary suppliers of the software and services for mainframe computers well into the 1960s were the manufacturers of these machines. In the case of IBM, which leased rather than sold many of its machines, the costs of software and services were "bundled" with the lease payments. By the late 1950s, however, a number of independent firms had entered the custom software industry. These firms included the Computer Usage Company and Computer Sciences Corporation, both of which were founded by former IBM employees (Campbell-Kelly 1995). In the late 1950s, the Computer Usage Company secured contracts with National Aeronautics and Space Administration (NASA) and had a successful initial public offering of shares. Many more independent firms entered the mainframe software industry during the 1960s.

Procurement of products and services by the federal government was an important factor in the early development of the software industry. IBM was the primary supplier of computers for the SAGE air defense project, but the RAND Corporation was the contractor responsible for the bulk of the huge amount of software required for SAGE. RAND, in turn, created a Software Development Division to produce the software. This division separated from RAND in 1956, forming the Systems Development Corporation. Because large-scale software development projects of this sort were well beyond the technological or scientific "frontier" of academic computer science (a discipline that itself scarcely existed in the early 1950s), the SAGE software development project acted as a "university" of sorts for hundreds of software programmers, laying the foundations for the software industry's future development within the United States (Campbell-Kelly 1995). To facilitate this training role, and in part because the Systems Development Corporation was restricted by Air Force pay scales, the company encouraged turnover of employees. The "SAGE alumni," in turn, contributed to the development of the broader software industry (Langlois and Mowery 1996). For example, one such programmer noted in the early 1980s, "the chances are reasonably high that on a large data processing job in the 1970s you would find at least one person who had worked with the SAGE system" (Benington 1983, 351).

In the late 1950s and early 1960s, defense contractors, including TRW, MITRE Corporation, and Hughes, began to produce large-scale systems software for military applications under federal contracts. IBM and other mainframe computer manufacturers also produced one-of-a-kind software applications for customers and became important suppliers in the software-contracting industry. Much of the software-related knowhow developed from defense contracts and the Apollo manned space flight program "spilled over" to commercial applications. For example, IBM's collaboration with American Airlines to develop the SABRE reservation system drew upon IBM's background with the SAGE development program (Campbell-Kelly 1995).

Federal procurement programs influenced the evolution of specific programming languages as well. A Department of Defense effort to establish a standard programming language resulted in the widely used "common business-oriented language," COBOL. The Department of Defense required that general-purpose computers purchased by the military support COBOL and that any business-related applications for defense programs be written in the language. Because the Department of Defense accounted for such a large share of the market for custom applications software, its procurement requirements facilitated the development and diffusion of COBOL (Flamm 1987).

The second era in the software industry's development—roughly from 1965 to 1978—witnessed significant entry by independent producers of standard software. Although independent suppliers of software had begun to enter the industry by the 1960s in the United States, computer manufacturers and users remained important sources of both custom and standard software during this period. Some service bureaus that had provided users with operating services and programming solutions began to unbundle the pricing of their services from software sales, providing yet another cohort of entrants into the independent development and sale of traded software. Sophisticated users of computer systems, especially users of mainframe computers, also developed expertise in creating solutions for their applications and operating system needs. A number of leading U.S. suppliers of commercial software were founded by computer specialists formerly employed by major mainframe users.

Steinmueller (1996) argues that three developments contributed to the expansion of the independent software industry in the United States during the 1960s. First, IBM's introduction of the 360 in 1965 provided a single mainframe architecture that utilized a standard operating system spanning all machines in this product family. This development increased the installed base of mainframe computers that could use packaged software designed to operate specific applications, and it made entry by independent developers more attractive. Second, IBM "unbundled" its pricing and supply of software and services in 1968, a decision that was encouraged by the threat of antitrust prosecution.[13] The "unbundling" of its software by the dominant

13. As the U.S. International Trade Commission (1995) points out, U.S. government procurement of computer services from independent suppliers aided the growth of a sizeable population of such firms by the late 1960s. These firms were among the first providers of custom software for mainframe computers after IBM's unbundling of services and software.

manufacturer of hardware (a firm that remains among the leading software suppliers worldwide) provided opportunities for the growth of independent software vendors. Third, the introduction of the minicomputer in the mid-1960s by firms that typically did not provide "bundled" software and services opened up another market segment for independent software vendors.

During the late 1970s and 1980s, the development and diffusion of the desktop computer produced explosive growth in the traded software industry. Once again, the United States was the "first mover" in this transformation and quickly emerged as the largest single market for packaged software. Rapid adoption of the desktop computer in the United States supported the early emergence of a few "dominant designs" in desktop computer architecture, creating the first mass market for packaged software. The independent software vendors (ISVs) that entered during this period were largely new to the industry. Few of the major suppliers of desktop software came from the ranks of the leading independent producers of mainframe and minicomputer software.

The large size of the U.S. packaged software market, as well as the fact that it was the first large market to experience rapid growth (reflecting the earlier appearance and rapid diffusion of mainframe and minicomputers, followed by the rapid growth of desktop computer use during the 1980s), gave the U.S. firms that pioneered in their domestic packaged software market a formidable "first-mover" advantage that they exploited internationally. During the 1990s, U.S. firms' market shares in their home market exceeded 80 percent in most classes of packaged software and exceeded 65 percent in non-U.S. markets for all but "applications" software.[14]

Much of the rapid growth in custom software firms during the 1970s reflected expansion in federal demand, which, in turn, was dominated by Department of Defense demand. But just as had been the case in the semiconductor industry, defense markets gradually were outstripped by commercial markets although this trend occurred more gradually in software than in hardware or semiconductors. There exists no reliable time series of Department of Defense expenditures on software procurement that employs a consistent definition of software (e.g., separating embedded software from custom applications or operating systems and packaged software). Nevertheless, the available, imperfect data suggest that in constant-dollar terms, Department of Defense expenditures on software increased more than thirtyfold between 1964 and 1990 (figure 5.7; see also Mowery and Langlois 1996). Throughout this period, Department of Defense software demand was dominated by custom software, and Department of Defense and federal government markets for custom software accounted for a substantial share

---

14. Most analyses of packaged-software markets distinguish among "operating systems" (the software used to control the operations of a given desktop, mainframe, or minicomputer), "applications" (software designed to support specific, generic functions such as word processing or spreadsheets), and "development tools" (such as programming languages, application development programs). For further details, see U.S. Department of Commerce (2000).

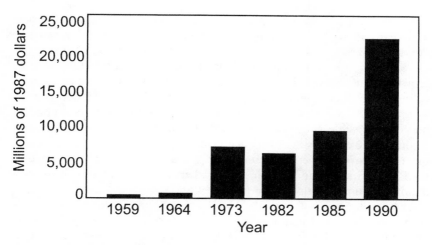

Fig. 5.7   U.S. Department of Defense software procurement, 1959–1990
Source: Langlois and Mowery (1996, 69, fig. 3.4). By permission of Oxford University Press, Inc.

of total revenues in this segment of the U.S. software industry. But despite this increase in Department of Defense demand, by the early 1990s, the Department of Defense accounted for a declining share of the U.S. software industry's revenues.

This declining share of meant that the defense market no longer exerted sufficient influence on the path of R&D and product development to benefit from generic academic research and product development—defense and commercial needs had diverged. The tangled history of the Department of Defense's "generic" software language, Ada, unveiled in 1984, illustrates the declining influence of federal procurement on the rapidly growing software industry. Billed as a solution to the problems of system maintenance and software development resulting from the bewildering variety of software languages in use within defense systems, Ada was designed to be employed in all defense applications.

Ada proponents argued that by standardizing all Department of Defense programs around a single language, the commercial developers that no longer served the military market would be motivated to produce software that could be used in both civilian and military applications. But these aspirations were largely unrealized. Partly because of the huge difficulties associated with "inserting" Ada into the enormous "installed base" of defense-related software, the language failed to attract the attention of commercial developers. The contrast between the failure of this Department of Defense-supported language to take hold and COBOL's rapid diffusion into military and commercial applications highlights the tendency for the influence of defense-related R&D and procurement demand on the overall trajectory of innovation in a given technology to decline as commercial markets expand.

The fourth era in the development of the software industry, which began in the early 1990s, has been dominated by the growth of networking among desktop computers both within enterprises through local area networks linked to a server and among millions of users through the Internet. The growth of the Internet has facilitated the development of "open source" software. User-active innovation in software is hardly new, and the exchange by users of "shareware" also has a long history; but the Internet supports modifications in open source software without the creation of competing, incompatible versions. As with the previous eras of this industry's development, the growth of network users and applications has been more rapid in the United States than in other industrial economies, and U.S. firms have maintained dominant positions in these markets.

The development of both the U.S. computer hardware and computer software industries rested on a research and personnel infrastructure supported by federal R&D investments. Perhaps the most important result of these investments was the development of a large university-based research complex that provided a steady stream of new ideas, some new products, and a large number of entrepreneurs and engineers eager to participate in this industry. Like postwar defense-related funding of R&D and procurement in semiconductors, federal policy toward the software industry was motivated mainly by national security concerns; nevertheless, federal financial support for a broad-based research infrastructure proved quite effective in spawning a vigorous civilian industry.

As in other segments of the IT industry, U.S. antitrust policy played an important role in the development of the software industry. The unbundling of software from hardware was almost certainly hastened by the threat of antitrust action against IBM in the late 1960s. Moreover, as noted earlier, many of the independent vendors that responded to the opportunities created by the new IBM policy had been suppliers of computer services to federal government agencies. In addition, the relatively liberal U.S. policy toward imports of computer hardware and components supported rapid declines in price-performance ratios in most areas of computer hardware and thereby accelerated domestic adoption of the hardware platforms that provided the mass markets for software producers. By comparison, Western European and Japanese governments' protection of their regional hardware industries was associated with higher hardware costs and slower rates of domestic adoption, ultimately impeding the growth of their domestic software markets.

5.4   The Development of Human Capital

One of the most important contributions of federal R&D investments in IT was the creation of a substantial pool of scientists and engineers who contributed to innovation in IT and related technologies across a broad array of applications. As the preceding discussion of computer hardware and software emphasizes, federal investments in university research supported the creation of a new academic discipline, computer science. The number of computer science departments in U.S. and Canadian universities that were granting PhD degrees grew from 6 in 1965 to 56 by 1975 and to 148 in 1995 (Andrews, 1997, cited in National Research Council 1999). In other U.S. research universities, computer science PhDs were awarded by established electrical engineering departments, so these figures understate somewhat the growth in advanced degree programs in computer science. The number of computer science bachelor's degrees awarded by U.S. and Canadian universities grew from 89 in 1966 to 42,000 by 1986.

Although the 1966 figures almost certainly underestimate the actual number of undergraduate degrees in this field (because many electrical engineering degrees awarded in 1966 covered virtually identical coursework and research), the growth in training of computer science personnel is remarkable. Master's degrees in computer science grew at an average annual rate of more than 14 percent during 1966 to 1995, and PhD degrees increased from 19 in 1966 to more than 900 by 1995 (National Research Council 1999). Electrical engineering undergraduate degree production also grew during the 1966 to 1986 period, more than doubling from 11,000 to 27,000.

Federal research funds played a central role in supporting this expanded production of electrical engineering and computer science degree holders. The National Research Council's 1999 study of federal R&D in IT estimates that the share of graduate students in U.S. universities' computer science and electrical engineering departments supported by federal funds through fellowships, teaching assistantships, or research assistantships rose from 14 to 20 percent during 1985 to 1996 and that more than one-half of this financial support was provided in the form of research assistantships. The contributions of federal funds to degree production in the leading U.S. research universities were even higher, according to the 1999 study. Between 1985 and 1996, roughly 56 percent of graduate students in the computer science and electrical engineering departments at MIT; Carnegie-Mellon University; and the University of California Berkeley were supported entirely or partly by federal funds, with 46 percent of these students being supported by federally funded research assistantships. Twenty-seven percent of graduate students at Stanford University's electrical engineering and computer science departments received federal funds, and 50 to 60 percent of the PhD students enrolled in these departments were supported in whole or in part by federal funds.

5.5   Conclusions

This summary of technological and industrial development within semiconductor, computer hardware, and computer software highlights the important and constructive role played by federal policy. One of the most striking characteristics of federal policy was the fact that it extended well beyond the "conventional" technology policy tool of R&D spending to include military procurement, intellectual property rights, and antitrust policies. Although these various dimensions were not coordinated or formulated with any coherent strategy in mind, they had the effect of supporting both the "supply" of knowledge, know-how, trained personnel, and the "demand" for adoption of the technologies emerging from the R&D process.

Federal policies also consistently supported a striking degree of competition among R&D funders, competition among R&D performers (including competition among R&D performers for the award of R&D and procurement contracts), and competition among the various actors seeking to commercialize new applications. An important factor contributing to competition among R&D performers was the reliance on extramural R&D performers in many federal R&D programs in the IT sector (the same could be said of federal R&D spending in biomedical science, another area of highly productive public R&D investments). United States research universities in particular proved to be effective sources of both basic knowledge and technological advances, as well as skilled engineers, scientists, and entrepreneurs. Much of the effectiveness of these institutions rested on the intense interinstitutional competition for resources, prestige, students, and faculty that characterizes the U.S. higher education system. Interinstitutional competition motivated university administrators and faculty to seek resources from both the federal government and industry, a competitive dynamic that proved to be highly successful in generating technical and commercial advances.

This competitive environment for R&D performers and commercializers also benefited from a tough federal antitrust policy in the IT sector. Leading firms in both computers and semiconductors licensed their technological portfolios more widely and at lower cost and may have avoided pursuing infringers of their intellectual property because of federal antitrust oversight. Antitrust policy, as well as federal R&D and procurement policies, reinforced and contributed to an environment of relatively weak intellectual property rights in these industries for much of their early years. Interfirm knowledge flows, entry by new firms, and experimentation with new approaches to commercial applications all were almost certainly more significant in this environment of relatively weak formal intellectual property rights that would have been true in a "strong patent" environment. The merits of strong, broad patent protection thus must be considered with care and some skepticism in the early, formative years of a new technology-based industry (See Mowery and Nelson [1999] for further discussion).

The effectiveness of federal R&D spending in the information technology sector also was enhanced by the sheer scale of its R&D and related procurement efforts, reflecting the positive externalities in innovation that flow from a large installed base in these technologies. As Bresnahan and Greenstein

(1996) have pointed out, much of the innovation process in IT involves "co-invention" on the part of users as well as technology suppliers. The importance of this co-invention means that large-scale deployment of a given new technology can spark more user-driven experimentation, thereby accelerating refinement, innovation, and improvement. The history of the packaged software industry in the United States clearly indicates the advantages of scale. Similar advantages may well inhere in large-scale experiments with either a technological "infrastructure" or small-scale technologies in other sectors, including energy.

Federal policy in information technology complemented support for R&D with support for adoption in the early years of the development of semiconductors and computers, where federal procurement contracts proved to be as influential as federal R&D funding of industrial firms in supporting innovation and firm entry. Federal procurement was less salient in computer software although early federal contracts for computer support and custom software development aided the growth of at least some firms that subsequently became important suppliers of mainframe software.

The commercial development and eventual adoption of semiconductor and computer hardware technologies benefited significantly from the large military purchases of early versions of these technologies for national-security applications. In addition, the structure of these procurement programs often encouraged entry by new firms and significant technology flows among firms. In contrast, U.S. energy-R&D programs have tended to combine instability in R&D funding with little systematic effort to support demand for early versions of new technologies (National Research Council 2001).

In short, federal R&D and other policies were of great importance to the development of economically vibrant semiconductor and computer hardware and software industries that literally did not exist sixty years ago. Of course, the historical structure of these federal policies differs in some important respects (notably, in intellectual property rights) from their current posture in energy and other sectors. Moreover, the influence of public R&D and procurement efforts in IT waned as the technologies underpinning this sector matured. The "lessons" of federal policy toward IT innovation accordingly must be applied to other sectors, such as innovation directed toward solutions to global climate change, with considerable discrimination and caution.

References

Alic, John A., Lewis M. Branscomb, Harvey Brooks, Ashton B. Carter, and Gerald L. Epstein. 1992. *Beyond spinoff: Military and commercial technologies in a changing world.* Boston: Harvard Business School Press.

Business and Defense Services Administration. 1960. *Electronic component: Production and related data, 1952–1959.* Washington, DC: U.S. Department of Commerce.

Benington, Herbert D. 1983. Production of large computer programs. *Annals of the History of Computing* 5 (4): 350–61.

Borrus, Michael, James E. Millstein, and John Zysman. 1983. Trade and development in the semiconductor industry: Japanese challenge and American response. In *American industry in international competition,* ed. John Zysman and Laura Tyson, 142–248. Ithaca, NY: Cornell University Press.

Braun, Ernest, and Stuart MacDonald. 1982. *Revolution in miniature.* 2nd ed. New York: Cambridge University Press.

Bresnahan, Timothy F., and Shane Greenstein. 1996. Technical progress and co-invention in computing and in the uses of computers. *Brookings Papers on Economic Activity, Microeconomics:* 1–77.

Burrows, Peter. 1992. Bill Spencer struggles to reform SEMATECH. *Electronic Business,* May 18.

Campbell-Kelly, Martin. 1995. Development and structure of the international software industry, 1950–1990. *Business and Economic History* 24 (2): 73–110.

Chandler, Alfred D., Jr. 2001. *Inventing the electronic century.* New York: Free Press.

Clement, J. R. B. 1987. Computer science and engineering support in the FY 1988 budget. In *AAAS report XII: Research & development, FY 1988,* ed. Intersociety Working Group. Washington, DC: American Association for the Advancement of Science.

———. 1989. Computer science and engineering support in the FY 1990 budget. In *AAAS report XIV: Research & development, FY 1990,* ed. Intersociety Working Group. Washington, DC: American Association for the Advancement of Science.

Clement, J. R. B. and D. Edgar. 1988. Computer science and engineering support in the FY 1989 budget. In *AAAS report XIII: Research & development, FY 1989,* ed. Intersociety Working Group. Washington, DC: American Association for the Advancement of Science.

Electronic Industries Association. 1969. *Electronic industries yearbook.* Washington, DC: Electronic Industries Association.

Fabrizio, Kira R., and David C. Mowery. 2007. The federal role in financing major innovations: Information technology during the postwar period. In *Financing innovation in the United States, 1870 to the present,* ed. Naomi R. Lamoreaux and Kenneth L. Sokoloff, 283–316. Cambridge, MA: MIT Press.

Fisher, Franklin M., James W. McKie, and Richard B. Mancke. 1983. *IBM and the U.S. data processing industry.* New York: Praeger.

Flamm, Kenneth. 1987. *Targeting the computer.* Washington, DC: Brookings Institution.

———. 1988. *Creating the computer.* Washington, DC: Brookings Institution.

Goldstine, Herman H. 1993. *The computer from Pascal to Von Neumann.* 2nd ed. Princeton, NJ: Princeton University Press.

Grindley, Peter, David C. Mowery, and Brian Silverman. 1994. SEMATECH and collaborative research: Lessons in the design of high-technology consortia. *Journal of Policy Analysis and Management* 13 (4): 723–58.

Holbrook, Daniel, Wesley M. Cohen, David A. Hounshell, and Steven Klepper. 2000. The nature, sources, and consequences of firm differences in the early history of the semiconductor industry. *Strategic Management Journal* 21 (10–11): 1017–41.

Klepper, Steven. 2009. Silicon Valley—A chip off the old Detroit bloc. In *Entrepre-*

neurship, growth, and public policy, ed. Zoltan J. Acs, David B. Audretsch, and Robert J. Strom, 79–115. New York: Cambridge University Press.

Langlois, Richard N., and David C. Mowery. 1996. The federal government role in the development of the U.S. software industry. Chap. 3 in *The international computer software industry: A comparative study of industry evolution and structure,* ed. David C. Mowery, 53–85. New York: Oxford University Press.

Langlois, Richard N., and W. Edward Steinmueller. 1999. The evolution of competitive advantage in the worldwide semiconductor industry, 1947–1996. In *Sources of industrial leadership: Studies of seven industries,* ed. David C. Mowery and Richard R. Nelson, 19–78. New York: Cambridge University Press.

Lécuyer, Christophe. 2005. *Making Silicon Valley: Innovation and the growth of high tech, 1930–1970.* Cambridge, MA: MIT Press.

Levin, Richard C. 1982. The Semiconductor Industry. In *Government and technical progress: A cross-industry analysis,* ed. Richard R. Nelson, 9–100. New York: Pergamon.

Macher, Jeffrey T., David C. Mowery, and David A. Hodges. 1999. Semiconductors. In *U.S. industry in 2000,* ed. David C. Mowery, 245–85. Washington, DC: National Academies Press.

Malerba, Franco. 1985. *The semiconductor business: The economics of rapid growth and decline.* Madison, WI: University of Wisconsin Press.

Misa, Thomas J. 1985. Military needs, commercial realities, and the development of the transistor, 1948–1958. In *Military enterprise and technological change,* ed. Merritt Roe Smith, 253–87. Cambridge, MA: MIT Press.

Mowery, David C. 1999. The computer software industry. In *Sources of industrial leadership: Studies of seven industries,* ed. David C. Mowery and Richard R. Nelson, 133–68. New York: Cambridge University Press.

Mowery, David C., and Richard R. Langlois. 1996. Spinning off and spinning on (?): The federal government role in the development of the U.S. computer software industry. *Research Policy* 25 (6): 947–66.

Mowery, David C., and Richard R. Nelson. 1999. Conclusion: Explaining industrial leadership. In *Sources of industrial leadership: Studies of seven industries,* ed. David C. Mowery and Richard R. Nelson, 1–18. New York: Cambridge University Press.

Mowery, David C., and Nathan Rosenberg. 1999. *Paths of innovation: Technological change in 20th century America.* New York: Cambridge University Press.

National Research Council. 1999. *Funding a revolution: Government support for computing research.* Washington, DC: National Academies Press.

———. 2001. *Energy research at DOE.* Washington, DC: National Academies Press.

National Science Foundation (NSF). 1998. *Federal funds survey, fields of science and engineering research historical tables, fiscal years 1970–1998.* Washington, DC: NSF.

Redmond, Kent C., and Thomas M. Smith. 1980. *Project whirlwind: History of a pioneer computer.* Bedford, MA: Digital Press.

Riordan, Michael, and Lillian Hoddeson. 1997. *Crystal fire: The birth of the information age.* New York: Norton.

Steinmueller, W. Edward. 1996. The U.S. software industry: An analysis and interpretive history. In *The international computer software industry: A comparative study of industry evolution and structure,* ed. David C. Mowery, 15–52. New York: Oxford University Press.

Tilton, John E. 1971. *International diffusion of technology: The case of semiconductors.* Washington, DC: Brookings Institution.

U.S. Department of Commerce. 1960. *Electronic component: Production and related data, 1952–1959.* Washington, DC: Business and Defense Services Administration (BDSA), U.S. Department of Commerce.

———. 1960–1968. Consolidated tabulation: Shipments of selected electronic components. Washington, DC: Business and Defense Services Administration (BDSA). Mimeograph.

———. 1969–1977. Shipments of defense-oriented industries. Current Industrial Reports, Series MA-175. Washington, DC: U.S. Bureau of the Census.

———. 2000. *U.S. industry and trade outlook 2000.* Washington, DC: GPO.

U.S. Department of Defense, Defense Science Board (DSB). 1987. *Report of the DSB Task Force on Defense Semiconductor Dependency.* Washington, DC: U.S. Department of Defense.

U.S. Department of Defense, Joint Service Task Force on Software Problems. 1982. *Report of the DOD Joint Service Task Force on Software Problems.* Washington, DC: U.S. Department of Defense.

U.S. International Trade Commission. 1995. *A competitive assessment of the U.S. computer software industry.* Washington, DC: U.S. International Trade Commission.

Usselman, Steven W. 1993. IBM and its imitators: Organizational capabilities and the emergence of the international computer industry. *Business and Economic History* 22 (2): 1–35.

Von Neumann, John. [1945] 1981. First draft of a report on the EDVAC, 1945. Reprinted in *From ENIAC to UNIVAC,* ed. N. Stern, 177–246. Bedford, MA: Digital Press.

6

Nurturing the Accumulation of Innovations
Lessons from the Internet

Shane Greenstein

As the Internet diffused throughout the 1990s, it touched a wide breadth of economic activities. The diffusion transformed the use of information technology throughout the economy. It led to improvements in products, lower prices, the development of new capabilities, and the development of many innovations that enabled productivity improvements among business users. It diffused to the majority of homes and businesses, altering the way people shop, research, play, and relate socially.

The Internet began as a government sponsored operation in the 1970s and 1980s and grew into a commercial industry in the 1990s. At first, the Internet lacked market-oriented focusing devices or economic inducement mechanisms typically associated with directing efforts toward the most valuable innovative outcomes.[1] There were contracts for carrier services between government buyers and commercial suppliers, for example, but no general market orientation toward the pricing of the exchange of traffic between carriers. There also were a few providers of Internet equipment for government users but no waves of inventive entrepreneurial entry. There were managers who understood the specific needs of their niche user communities but no

Shane Greenstein is the Elinor and Wendell Hobbs Professor of Management and Strategy at the Kellogg School of Management, Northwestern University, and a research associate of the National Bureau of Economic Research.

I thank Guy Arie, Tim Bresnahan, David Clark, David Crocker, Rebecca Henderson, Franco Malerba, Richard Newell, Bonnie Nevel, John Quarterman, Craig Partridge, Richard Schmalensee, Alicia Shems, Scott Stern, and Stephen Wolff for extraordinarily useful conversations and comments. This chapter is part of a larger project funded by the Kaufman Foundation, the Searle Center at Northwestern University, and the dean's office at the Kellogg School of Management at Northwestern. I am grateful for the funding. I am responsible for all remaining errors.

1. For more on focusing devices in general, see, for example, Rosenberg (1977).

possibility for tailoring new products and services to every potential new set of users.

How could an institutional setting that lacked market-orientation yield a set of innovations that supported the creation of massive market value only a few years later? This chapter helps explain the progression. The chapter divides the Internet's development into a precommercial and commercial era. The precommercial period encompasses the 1970s and 1980s and some of the early 1990s, when the government controlled the research and development of the Internet and its components. The commercial period, which arose after the government opened control of the network to commercial interests, encompasses the mid-1990s and onward to present day.

These two eras illustrate two distinct models for accumulating innovations over the long haul. The precommercial era illustrates the operation of several useful nonmarket institutional arrangements. It also illustrates a potential drawback to government sponsorship—in this instance, truncation of exploratory activity. The commercial era illustrates a rather different set of lessons. It highlights the extraordinary power of market-oriented and widely distributed investment and adoption, which illustrates the power of market experimentation to foster innovative activity. It also illustrates a few of the conditions necessary to unleash value creation from such accumulated lessons, such as standards development and competition, and nurturing legal and regulatory policies.

6.1   The Precommercial Internet under DARPA

It may be tempting to compare the Internet to historically archetypical big inventions sponsored by government, such as the Manhattan and the Apollo projects.[2] However, these archetypes for developing technical breakthroughs are not good models for understanding what happened during the creation of the Internet. The Internet was not a single urgent project in a single lab devoted to engineering a single object. In fact, the early development of the Internet is rather less exciting than commonly assumed.

The Internet began as a government-sponsored project with a restricted set of users and uses. It occurred at a time when theory pointed in a few promising directions, but nobody, not even the experts, knew where implementation would lead. The project involved a vastly dispersed set of technically adept participants with a shared interest in the project but otherwise heterogeneous needs and outlooks. The Internet developed slowly, and through a rather mundane process, accumulating capabilities over time from a vast number of contributors.

The Internet's early development fit into an archetype called "collective invention." Collective invention is "a process in which improvements

2. Much of the material in this first section summarizes Greenstein (2010a).

or experimental findings about a production process or tool are regularly shared."[3] There was no single user who demanded a technology such as the Internet, nor any single inventor for it. Rather, five partially overlapping groups played a role in shaping attributes that each valued, with an accumulation of innovative contributions over time. The first two groups were the primary decision makers at funding agencies: the Department of Defense (DOD) and the National Science Foundation (NSF). The other three were programmers/inventors, administrators, and application users. Many were funded by the government agencies and given considerable discretion. Others became participants over time and added their own contributions within their own budgetary limitations.

The U.S. military budget served as the first source of funds for the precommercial Internet, while the NSF largely served as the source of funds from 1986 to the end of government involvement in 1995. At a very basic level, the U.S. government paid for much of the research and development (R&D) of the Internet during this period, and the government was the organization behind the early "demand" for the breakthrough technical achievements that became recognized as the Internet.

This effort did not begin like a military procurement project, as if the Internet were a military rocket procured from several suppliers. In such a procurement process, a group of expert U.S. military personnel consider and deliberate at great length with great foresight about the needs of the government, and then they issue a set of specifications for a product or service with a predetermined set of attributes. However, in the case of the Internet, the DOD's Department of Advanced Research Projects Agency (DARPA) funded the core research and development work that led to the Internet—indeed, this work was but one of many DARPA projects on the frontiers of computer science.[4] The project first intended to build prototypes for a packet-switching data-communications network of networks, which pushed the boundaries of computing at the time. Both "packet-switching" and "a network of networks" were budding theoretic concepts. The Department of Advanced Research Project Agency's administrators wanted innovations in the form of ideas, new designs, and new software. The administrators also desired all of these innovations be portable to military operations in the long run, as required under the Mansfield Amendment (stipulating that DOD funding be relevant to military's mission). While the demand for these innovative solutions was quite general, the specifics were undetermined. The U.S. military faced issues with its own computing facilities and operations

3. See Allen (1983) for the introduction and illustration of the concept. The quote is from Meyer (2003, 4), who provides further extensions to modern examples.

4. This organization was originally founded as the Advanced Research Projects Agency (ARPA) and became DARPA in 1972. For the sake of simplicity, I use only the name DARPA throughout this chapter. See Norberg, O'Neill, and Freedman (1996) and Roland and Shiman (2002).

that justified the R&D expense, even though its own managers could not concretely describe the object that would result.[5]

The military sought a robust design for a communications network, and its potential value was self-evident. Keeping communications functioning in spite of a blown/cut line, for example, has military value in hostile battlefield conditions. Packet switching held the promise to achieve this attribute by allowing data to flow along multiple paths, unlike a circuit-switched telephone network in which calls follow a preset path programmed into central office telephone switches. In principle, an inexpensive packet switching network could also cover vast geographic distances, which could support the sharing of expensive computing resources over such distances. That too had self-evident military value. For example, military users in many locations—even potentially dangerous locations—could access databases housed in another (potentially safer) location.

An additional technical and pragmatic aspiration also played a role. An ideal network of networks could facilitate the movement of data between computer systems—mostly mainframes in this era—that otherwise could not interoperate seamlessly. A system that could enable the exchange of data and communication between computing systems without frequent human intervention would save the military time and personnel expenses and help realize new strategic capabilities. Coordinating the exchange, combination, and filtering of data between computer systems generated numerous logistical and organizational problems for military operations that increasingly depended on computing.

Although these innovations later would be portable to nondefense uses, that was not among the relevant criteria for the program at the outset. Initially, there was no explicit requirement that the innovation work with all (or even most) computer systems in nonmilitary uses, though that was a likely by-product because nonmilitary uses of computing overlapped in some applications and functionality with military uses. However, the program was informed by a general understanding that shaped all activities in information technology (IT) within DARPA: having a healthy U.S. information technology sector was a valuable military advantage in the long run.

Eventually several prototypes for a packet switching network were engineered. With additional work, these innovative designs turned into a prototype of an operating network, operated by managers from Bolt Beranek and Newman (BBN), a research contractor that subcontracted through DARPA. A number of researchers and their students became familiar with its design and operational principles. The network grew from there, covering

5. Norberg, O'Neill, and Freedman (1996) repeatedly stress that DARPA's funding of packet switching research in the 1960s and 1970s met concerns about whether the funding was relevant to military mission, as required by the Mansfield Amendment, which was proposed several times and eventually passed in 1973. The research anticipated enhancing the "command and control" capabilities of commanders increasingly reliant on their computing resources.

more locations and more participants throughout the 1970s. The network extended into many research laboratories and universities with funding from DARPA.

6.1.1 A Skunk Works and Wild Ducks

From the outset when DARPA's "packet switching" and "network of networks" project began, the desired attributes for the Internet represented a radical technological departure from existing practice. To understand the development of radical technologies within the military, it is best to begin with an understanding of two terms: a *skunk works* and *wild ducks*.

A *skunk works* is an organizational home for frontier development projects.[6] It is housed away from the main operations of an organization, sometimes in secret or with organizational barriers, and often with top management support for these barriers. Typically, the development projects involve something of value to the future of the organization but are not directly connected to its present operational or service missions. In this case, DARPA itself is the military's skunk works. The mission of the agency is research oriented, not operational, although in the case of the Internet, operational issues eventually became salient as well. Broadly construed, DARPA's mission is to develop radical new concepts and operations to transform military operations through development of new technologies. The agency had been established after Sputnik, and it was deliberately not beholden to the short-run operational needs of any of the armed services, although its innovations were required to eventually enhance some military function.

Wild ducks are a particular group of technically adept and innovative contributors, often considered social outsiders by those controlling funding. Wild ducks can encompass a range of behaviors and social differences that are regarded as potentially disruptive and costly to the regular operations of an organization. The reverse also often holds—that is, wild ducks often regard the practices of those involved in regular operations as interfering with their inventive activity. During DARPA's research into what became the Internet, the inventive individuals behind it were largely wild ducks.

Wild ducks had been a colloquial term of art in computing for decades, coined by IBM Chief Executive Officer (CEO) Thomas Watson, Jr. to describe the innovative practices of his technical team.[7] At IBM, wild ducks

---

6. The phrase originated from a project for the Air Force at a division of Lockheed Martin. The division had called itself the "Skonk Works" after a phrase from Al Capp's *Lil' Abner* cartoon—the skonk works was a "secret laboratory" that operated a backwoods still. The label became well known throughout the industry, in part because it was considered humorous and saucy. *Lil' Abner's* publisher eventually asked Lockheed Martin to change it, and "skunk works" emerged from there (Rich and Janus 1994).

7. Watson Sr. encouraged social conformity in his firm because he believed it made his sales force more effective (for example, all salesmen had to wear blue suits). But Watson Jr. came to a different understanding with his technical talent. His wild ducks had permission to be diverse, so long as they invented. For many stories related to wild ducks, see Maney (2003).

were uncompromising avant-garde thinkers in their field, often chasing visions they saw as aesthetic.[8] The value of their ideas could defy evaluation ex ante, and in many cases, it was hard to evaluate even after a prototype was developed. However, to fully realize the innovative potential of wild ducks and allow them to coexist within a mainstream organization, IBM separated the wild ducks from others, which kept valuable inventions temporarily hidden, unthreatening to others in the business, and, therefore, flowing until needed. IBM used wild ducks to develop innovative products and used the mainstream sales force to systematically and uniformly sell these products. IBM thereby kept control of the computing platform and ensured its commercial success by making sure it did not remain static.

The wild ducks arrangement worked well for the DOD. If the wild ducks failed to realize their grandest innovative vision, then almost nobody had to be bothered. If, on the other hand, the wild ducks invented something that others within the mainstream organization could appreciate and use, then the best scenario would be achieved. However, there was one troubling scenario: what would the mainstream organization do if the most valuable inventions could not be integrated into existing operations? In this case, the wild duck arrangement allowed the military to at least defer questions about a costly integration until the time when or if the innovation proved fruitful. If the need arose, then such knotty questions would have to be addressed, but not before then.

As it turned out, a particularly inventive group of wild ducks in the pre-commercial Internet accumulated a range of inventions, eventually bringing about a large economic gain for participants. More to the point, these innovations, which comprised the basic building blocks for the Internet, turned out to have enormous value when transferred to nonparticipants. Some of these inventors were established university researcher, such as Paul Baran,

---

8. "In IBM we frequently refer to our need for 'wild ducks.' The moral is drawn from a story by the Danish philosopher Soren Kierkegaard, who told of a man who fed the wild ducks flying south in great flocks each fall. After a while some of the ducks no longer bothered to fly south; they wintered in Denmark on what he fed them. In time they flew less and less. After three or four years they grew so lazy and fat that they found difficulty in flying at all. Kierkegaard drew his point: you can make wild ducks tame, but you can never make tame ducks wild again. One might also add that the duck who is tamed will never go anywhere any more. We are convinced that any business needs its wild ducks. And in IBM we try not to tame them" (Watson 1963, 27–28).

The phrase was well known and widely acknowledged. For example, see this recollection of a Thomas Watson, Jr. speech from a former IBM employee, published many years later in a letter to the New York Times in 1989. "I talk a lot around here about wild ducks, and people kid me a good deal about my wild ducks. But it takes a few wild ducks to make any business go, because if you don't have the fellows with the new ideas willing to buck the managerial trends and shock them into doing something new and better, if you don't have those kind of people, the business pretty well slows down. So I would tell a 21-year-old I.B.M.'er what I've told a lot of 21-year-old college people . . . that is, that the priceless ingredient that a youngster has when he starts in business is that sense of not compromising beyond a certain point" (http://www.nytimes.com/1989/05/07/business/1-wild-ducks-048289.html).

Joseph Licklider, or Leonard Kleinrock, and their reputations would be further enhanced by their involvement in designing packet-switching networks. Other researchers would become affiliated with the project right at the outset of their careers and remain involved throughout their careers. This included Steve and David Crocker, David Clark, John Postel, and Vint Cerf, among many others. The Internet Protocol Suite known as Transmission Control Protocol/Internet Protocol (TCP/IP) emerged from their efforts.[9] Their achievements would be recognized by contemporaries, and they gained reputations over time from those achievements.

The Department of Advanced Research Projects Agency's program for fostering innovations in computing departed from the archetype of a skunk works practiced among military contractors.[10] The continuity in DARPA's managerial procedures and policies borrowed considerably from practices for R&D and military procurement, melding them into a goal-oriented research and development project administered by technically capable program officers executing a general vision. One key departure involved the amount of discretion given program officers. Though they were reviewed eventually, in the short run, many had freedom to make the decisions they thought would work best. Another principal departure involved geography. The Department of Advanced Research Projects Agency's skunk works was not physically housed in a single organization in Washington, D.C. Instead, it was *administered* from D.C., but the work was geographically dispersed to many locations in research organizations and universities across the country. The Department of Advanced Research Projects Agency sent money for projects organized by key researchers, who maintained their laboratories at their own universities. Money was also sent to contracting research organizations, such as BBN (in Cambridge, Massachusetts), the RAND Corporation (in Santa Monica, California), and Stanford Research Institute (in Menlo Park, California). While DARPA provided funds to support labs, buy equipment, and pay graduate students at these locations, the government agency was able to take advantage of building on what was already there, both in terms of institutions and brainpower.

Dispersed geography mattered in several other ways. Innovative improvements arose and accumulated in different places, yielding a variety of lessons and insights at a time when theory pointed in many directions and implementations were scarce. Collectively, this program began accumulat-

---

9. Said simply, TCP determined a set of procedures for moving data across a network and what to do when problems arose. If there were errors or specific congestion issues, TCP contained procedures for retransmitting the data. While serving the same function of a postal envelope and address, IP also shaped the format of the message inside. It specified the address for the packet, its origin and destination, a few details about how the message format worked, and, in conjunction with routers, the likely path for the packets toward its destination. An extensive explanation of TCP/IP can be found in many publications. See, for example, Leiner et al. (2003) or Abbate (1999).

10. See Norberg, O'Neill, and Freedman (1996).

ing improvements and suggestions from a diversity of sources, which were loosely coupled to one another through their common funding source and, nontrivially, shared scientific and engineering goals.

Program officers encouraged this sharing through arranged face-to-face meetings and communications.[11] Despite the geographic dispersion, participants shared a sense of identity about the whole project, and they were encouraged to share innovations with one another. Inventors also were encouraged to pay close attention to how their meta-design facilitated inventive specialization across the entire program. In addition, participants developed norms for documentation to facilitate knowledge retention and improvements built on earlier advances.

Unlike typical project management, program officers in this case did not initially rely on some of the institutions typically affiliated with academic science, such as peer-review, formal proposals with multiple stages of review, and panels of reviewers. They did solicit research proposals on occasion, but not necessarily proposals promising specific incremental advances within short time horizons. Instead, the program officers often asked for short broad proposals, picked stars, made general agreements with them about the long-term goals, funded their labs with uncommonly large amounts of money (for the discipline at the time), and gave them large amounts of discretion to pursue those goals in the manner they saw fit. In exchange for this funding, the researchers were required to attempt technically ambitious projects, participate in certain conferences, document and share their results with each other, and contribute to the training of a new generation of researchers, among other things.

Large sums of money invested by the DOD sustained continuity in its operation and continued improvement. The level of funding is notable because no program officer ever asked for concrete invention on a specific time frame, for example, and most of the inventors would have considered meeting such requirements to be pointless and absurd bureaucratic milestones. At the same time, many program officers were technically sophisticated enough to follow specific advanced developments. Some of them even contributed inventions to the efforts. In fact, DOD program officers often did the evaluation themselves or with a small set of consultations and not necessarily using evaluation by peers.

6.1.2   Nurturing Useful Prototypes

Precisely because a skunk works seeks to break with established patterns to facilitate experimentation and protect it from the objections of other organizations or their parent entity, a skunk works faces numerous chal-

---

11. Building coherent scientific communities around nascent technologies was an explicit part of the mission of every program officer in this era. See, for example, Norberg, O'Neill, and Freedman (1996) and Roland and Shiman (2002).

lenges meeting existing user needs. Its challenges are even greater when the participants in the skunk works create inventions for needs that most potential users have not yet even recognized. In the case of the DARPA Internet project, however, innovation and operations began to overlap. As a result, instead of meeting bureaucratic requirements, inventors were held to a different test: they had to eat what they grew. That is, their innovations were put into use comparatively quickly. The overlap between operations and invention played a key role in fostering *useful* innovations.

The first and second generation of Internet researchers quickly became familiar with a second unusual feature of their skunk works: new ideas grew out of their own experiences and often stemmed from their own needs.[12] Because inventers were also users, they were motivated to develop working prototypes into operational pieces that they and others could employ. Working prototypes were crude models of innovation in need of refined improvement. Often oriented toward demonstrating the proof of a new concept, these were deliberate interim manifestations of proposals, aimed to explore and, if possible, solve a piece of a problem and to help the inventors learn. Through their own use, many of these inventors became interested in issues that moved beyond simply illustrating a concept with their prototypes. They were introduced to issues associated with refining and maintaining workable versions of their inventions in a functioning and operational network—and not just any network, but a network they all used.

In the short run, mixing inventive activities with operational activities had a very direct effect on orienting innovation. Although using a common network, each group of researchers began working in its own direction, with its own working prototypes, for its own use as well as use by others. Due to their common affiliation with DARPA and common use of the network (which became known as the DARPAnet), the researchers began to make their prototypes interoperate with each other.

Many analysts of computing markets today stress the importance of a "killer application"—an application so compelling it justifies complementary investments. Early Internet innovators quickly developed several killer applications—file transfer, (something close to what we today recognize as) instant messaging, and electronic communication that became electronic mail.[13] Arguably, electronic mail was not even the most central innovation of the skunk works, but it was one that every participant used. Its pragmatic

---

12. There is no clean line between generations. The first generation of Internet researchers grappled with engineering, creating the first packet-switching applications and prototypes, and demonstrating the viability of the concepts. The second generation contributed to the existing infrastructure, and, along with the first generation, built applications and scale (Crocker 2008a).

13. See Partridge (2008), Crocker (2008b), and the description in the Living Internet History sites (http://www.livinginternet.com/e/e.htm) for documentation of how subsequent technical improvements built on one another, beginning with an early project at the RAND Corporation in Los Angeles. These passages draw heavily from Partridge's and Crocker's accounts.

value was widely recognized among participants. More than fifty people made important contributions to the standard e-mail design in the 1970s and 1980s, and by the end of the decade, virtually all participants in the Internet made use of this design. Another lesson from the e-mail application innovation is that its usefulness was apparent at the time to the innovators but not to the sponsoring federal agency. As stated by Bob Kahn, DARPA "would never have funded a computer network in order to facilitate e-mail" because other goals were more paramount, and person-to-person communication over telephones appeared sufficient.[14]

The spread of e-mail highlights the essential paradox of a skunk works: protecting wild ducks leads to long-term benefits if the inventions get pointed in useful directions from an early stage. However, at an early stage, virtually nobody in an organization except the most technically sophisticated is able to assess whether the wild ducks have succeeded in moving in a useful direction or in achieving even the most basic milestones!

How did participants make such assessments then, particularly into the late 1970s, after the basic science was done but considerable room was left for implementing new improvements? The integration of innovations into immediate operation shaped the consensus about innovations and helped determine whether suggestions for new protocols merited attention. As improvements arose, those improvements became gradually embedded in routine processes. If installation administrators did not think the innovations useful, they did not get installed nor used. If they did get used, the inventions got refined and began to accumulate additional improvements.

One additional aspect of this experience deserves attention: the DARPA skunk works was a technical meritocracy. In a technical meritocracy, individuals advance in standing through commonly recognized technical achievements rather than by external credentialing. The technical meritocracy for the Internet survived as an informal consensus process in the 1970s and much of the 1980s. The meritocracy survived for several reasons. First, virtually all participants came out of an academic and research background. They found it natural to work within a technical meritocracy, developing consensus about improvements worth keeping. Second, most of the program officers shared this research background, and they justified their actions on a similar basis. Third, in any given area, the group of researchers and administrators tended to be small enough that a technical consensus could emerge comparatively quickly. Fourth, and crucially, the top managers in the DOD protected DARPA against other influences on decision making, such as promotion of researchers or their projects using criteria other than a technical meritocracy. That is significant when compared to the alternatives: promotion based on abject favoritism, outside political connections, seniority within university hierarchies, or fame acquired through prior accomplishments in other areas.

---

14. See Segaller (1998, 105).

Note, however, that this technical meritocracy was pragmatic in its orientation because innovation was put into use. If installation administrators did not find it useful, then the invention did not get used. If it did not get used, the invention did not get refined or accumulate additional improvements.

6.1.3   The Internet under the NSF

The example of the DARPA Internet project contains lessons of certain behavior to avoid. First and foremost, the DOD restricted participation in the use of the results from the DARPAnet experiments. These restrictions truncated the range of uses to which the technology could be put by truncating the set of users who could experiment with it. Administrators of DARPA partially recognized this limitation and eventually permitted its contractor BBN to spin off a division and start a packet-switching service in the early 1970s.[15]

The issue kept returning, however, in part because DARPA sponsored experiments that succeeded more than anticipated. By the end of the 1970s, the DARPAnet was operational, and though far from perfect, the key pieces of the engineering insights had moved far beyond their status as working prototypes. It connected a network of research contractors and university researchers who wanted to continue to collaborate with each other. The inventions were portable to others, who could (and did) independently design and operate their own networks. In fact, frustrations with gaining access to the DARPAnet motivated some participants to start their own networks. For example, both Bitnet and CSNet began in the early 1980s, partly as a response to restricted access to DARPANet. Both of these spin-off networks provided the functionality that users desired, enabling them to move data between computers in different locations, supporting file sharing, and enabling electronic mail. Each had a different architecture and rules for participation, however. CSNet aspired to provide connectivity only to computer science departments, while Bitnet connected computing systems between various researchers and universities. A third network at the time was more informally organized and went by the label UUCPNet or Usenet and involved numerous participants both from inside universities and outside.

The increasing growth of alternative networks showed that such connectivity interested numerous participants other than the military. The network had grown beyond the core concerns of the military. Eventually, more researchers wanted to participate than DARPA had an interest in supporting. The Department of Advanced Research Projects Agency also worried about compromising the security of its own network by allowing nonmilitary users to participate.

By the early 1980s, the limits of participation became a widely recog-

---

15. This company became known as Telenet, and grew into a very large commercial provider of packet-switching service. Eventually Telenet became part of the Sprint data network.

nized source of tension. Finally, in 1985, DARPA handed over control of part of the network to the NSF in order to open it to the many civilian researchers interested in using it.[16] By then, the community of innovators had evolved into a loose confederation of researchers from many locations, so this administrative change partly ratified what had already begun to happen informally.

Innovation under NSF funding differed in several respects from innovation under DAPRA funding. This is not surprising because the missions of the two organizations differ. The Department of Advanced Research Projects Agency is part of the DOD, while the NSF supports civilian research. Just as with DARPA, no requirement about an immediate civilian application shaped activities other than a general understanding that the NSF's needs could be met more easily and cheaply if the U.S. computing and communications industries remained healthy. And as with DARPA's motivation, aspirations for resource sharing shaped NSF's investment. Much investment was, therefore, aimed at packet switching and the creation of an electronic communication network among researchers. The packet switching would enable the movement of files between supercomputer centers and many universities. Supercomputers were expensive fixed investments with no geographic mobility. The NSF aimed to use the Internet to permit many researchers to connect with those supercomputers, making greater use of the capacity and sharing the huge computing power they embodied.

Another aspiration for the NSF concerned a scaling issue that DARPA had not yet faced. The U.S. research community increasingly took to using the communication network for file sharing and electronic communication, and throughout this period, traffic grew. The NSF aimed to build a routine and reliable network infrastructure, making it easy to adapt and spread to every place of higher learning in the United States—universities, community colleges, and research institutes.[17] Over time, the investment aimed to give a wide range of participants—students, faculty, and administrators— a taste for what the Internet could do to help them in their work, namely, transmit electronic communication, data files, news, and other types of messages over long distances. The goal required a system that would handle traffic of many orders of magnitudes greater than anything accomplished to that point.

However, as with DARPA, NSF's management also came with some restrictions on participation, thus perpetuating the limitations of experimentation—only users connected to civilian research institutions could make use of the NSFNET, not, for example, commercial interests (except those who supplied services to NSF). However, restrictions due to this "acceptable use policy" were less binding than they had been with DARPA, and

---

16. These issues are described in great detail in Abbate (1999).
17. See Frazier (1995) and Leiner et al. (2003).

for a few years, NSF's managerial control reduced many of the tensions in the research community over participation.[18]

The transfer to NSF had several more consequences. A new source of funding introduced a new budgetary process, a new outlook about the future, and a new set of priorities for operations. In particular, NSF managed the backbone of the network but gave discretion to many universities to modify their installations as they saw fit. The NSF also differed from DARPA in its more relaxed approach to outsourcing equipment supplies, which had later consequences for transitioning from NSF administration into wide commercial use. By the time the Internet was commercialized, the surrounding industry was already in place to meet the needs of the new commercial market. For example, by the early 1990s, there already was an industry building routers consistent with widely employed software protocols.

In the late 1980s, NSF presided over another seminal design choice—the switch to a routing protocol that allowed for more than one backbone.[19] Until the NSFNET came into existence, there was only one network and one backbone, and BBN operated it. The scale was limited, and, in contrast, the NSF anticipated supporting a much large network. The NSFNET, therefore, introduced additional backbones and regional carriers. In due time, the NSF worked with others to introduce routing protocols that no longer presumed the NSF would be the sole manager for the backbone. This was the beginning of the technical design changes necessary for evolution to a commercial Internet with multiple commercial carriers.

By the beginning of the transition to commercialization in the late 1980s, the Internet was a large-scale and reliable data communications network with a well-documented code base upon which any participant could build additional layers of applications. While no serious networking engineer thought the Internet's technical capabilities had stopped evolving, insiders generally acknowledged that the research-oriented Internet had matured, moving beyond its "nuts and bolts" stage of development.[20] At that point, Steve Wolff, then director of the NSFNET, recognized that there was no technical reason why the government had to solely operate the Internet. He also asserted that private firms could provide services as efficiently, or more so, than government-managed entities or subcontractors. He, therefore, ini-

---

18. The NSF's "acceptable use" policy restricted the use of the NSFNET to to any university research faculty, student, or institute that contributed to furthering the development of science in the United States.

19. The NSF switched from the routing protocol Exterior Gateway Protocol (EGP) and replaced it with Border Gate Protocol (BGP). The EGP protocol presumed a known pathway for connecting systems. The BGP protocol enables fully decentralized routing. To Internet veteran David Clark, making this change was one of the earliest technical signs of the pending arrival of commercial network and the retirement of NSFNET (David Clark (2008), personal communication between David Clark and the author, September 26, 2008).

20. This is the phrase used by Mandelbaum and Mandelbaum (1992). See also Leiner et al. (2003).

tiated a long series of steps (with the full support of the NSF's management) aimed at what would be a transfer of technology out of exclusive government management and use.

Wolff's decision in itself illustrates another extremely important lesson. When a technology reaches a point where private firms can commercialize it, the transfer does not necessarily happen on its own. It requires government managers who recognize this opportunity, and it may even require active nurturing from government officials, as it did in this case.

In the case of the Internet, this transition was quite early in some sense and quite late in another. By the time it was turned over to commercial use, the Internet had acquired most of the attributes that would lead to the transformation of every part of information and communications markets around the world. However, because of the NSF's "acceptable use" policy, there had been little experimentation with using the Internet for commerce. There also was little understanding about its cost structure outside of an academic environment. Few of the participants had incentives to fully explore how a wide range of interfirm procedures would accommodate pricing, such as how interconnected networks would settle payments for exchanging traffic.

All in all, the NSF's managers invested in numerous innovations that contributed to easing the transition to commercial markets. However, the limited experience with a variety of users undermined the ability of Wolff and his managers, as well as managers elsewhere, to forecast the appeal of new applications aimed at new commercial users.

6.1.4   The Cost of Innovation

It would be historically inaccurate to presume the funding for basic research about the Internet arose out of cost and benefit calculation designed to accelerate the arrival of those economic gains. The cost of the Internet was not of interest to the government, especially at the outset. The Department of Advanced Research Projects Agency quite explicitly did not use economic rationales to fund projects, and DARPA funded high-risk projects that "dealt with the development of fundamental or enabling technologies with which to reach newly defined DOD objectives. . . . When DARPA judged success, it applied technical rather economic measurement standards."[21] Likewise, the NSF invested in developing Internet technologies to meet its agency mission, not with the intent of producing large economic gains.

It is also not possible to perform a cost and benefit calculation with the benefit of hindsight. The total cost to the government of creating the Internet is difficult to ascertain. It is known that during NSF's management (approximately 1985 to 1995) the agency invested $200 million dollars in Internet technology.[22] However, this figure does not include the DARPA

---

21. See Norberg, O'Neill, and Freedman (1996, 7).
22. See Leiner et al. (2003).

funding that paid for most of the early invention in the 1970s and early 1980s. While DARPA's financial commitment to what became the Internet was undoubtedly considerable, to my knowledge, no historian of these events has made a precise estimate of its size.[23] In addition, the cost tally of the Internet is further complicated because both DARPA and the NSF relied on distributed investments—the agencies paid for investments in backbone facilities and facilities for data-exchange, but offered only minimal support for investing in installations at universities. Most universities invested heavily in their own computing facilities, paid for by university funds.

The cost of the Internet would also include the substantial number of failures that were part of DARPA's broad portfolio of investments in computing science more generally. For example, it would include DARPA funding for a range of computer science efforts that did not work out as well as planned, such as in artificial intelligence. It also does not include a range of other experiments in computer science that the NSF paid for and from which the general community of researchers learned.

The Internet also benefited from improvements in a wide range of computing equipment that would have occurred with or without government funding. Like any other IT-intensive activity, research on the Internet gained benefits from what was happening to all equipment based on advances in solid-state circuitry. It was easier to make innovative gains when many of the other complementary inputs into the effort improved at the same time.

In summary, the early Internet succeeded because of the mix of managerial wisdom, pragmatism, and technical meritocracy of those involved and because those players kept their efforts trained on scientifically worthwhile projects. The federal institutions sustained those efforts over a long period of time, building a community of researchers invested in innovating and refining attributes of the network. It eventually accumulated many attributes that today we recognize as the Internet and which today we recognize as valuable. None of this was easy, automatic, or necessarily inexpensive.

6.2   The Commercial Internet

Once commercialized, the Internet was unlike any commercial communications network that came before it. While it still could be described as a packet-switching network for moving data between computing clients (as had been envisioned from its inception), this description does not fully describe the early commercial form the Internet. It accumulated more capabilities and functions as a range of firms began to use pieces of it to enhance services provided to paying customers. Over time, "the Internet" became a label for not only the Internet but also for all the applications that accumu-

23. The entire expenditure for the IPTO, the agency within DARPA that funded most of the Internet, did not exceed approximately $500 million over its entire existence (1963 to 1986), and the funding for what became the Internet was but one of many IPTO projects (Norberg, O'Neill, and Freedman 1996).

Table 6.1   Revenue for access markets ($ millions)

	1998	1999	2000	2001	2002	2003	2004	2005	2006
Dial-up	5,499	8,966	12,345	13,751	14,093	14,173	14,081	12,240	10,983
DSL	n.a.	228	1,245	2,822	4,316	6,954	10,240	12,034	15,066
Cable	138	274	903	2,600	4,117	7,372	9,435	11,139	13,156
Wireless						n.a.	668	1,140	n.a.

Source: Greenstein and McDevitt (2009).
Note: n.a. = not available.

lated around the Internet, used pieces of the Internet, and commercialized new functions for the Internet. Together they delivered an impressive array of services to a wide range of users.

Supply of the Internet did not simply create its own demand. Rather, after years of development, a few applications were built that provided compelling value for tens of millions of decision makers. The size of the Internet access economy in the United States gives a sense of how big demand for the Internet became once it started to commercialize. For example, the revenue associated with providing Internet access is one of the largest categories of revenue out of the value chain for Internet services, and it grew quite large in only nine years (table 6.1).[24] By 2006, total revenues reached $39 billion; that is extraordinary for a technology that had almost no commercial service providers prior to 1989.[25]

These revenue levels are important to stress because access fees generated most of the revenue during the first decade of the commercial Internet. A typical U.S. household spent more than three-quarters of its online time at free or advertising-supported sites.[26] Although subscription-based services and advertising services started growing rapidly after 2003, the amount spent on access fees each year has far exceeded advertising revenue. For example, the $39 billion in access revenue in 2006 compared with $9.7 billion in Web Search Portal revenue (which includes advertising) and $12.8 billion

24. A value chain is a set of interrelated activities that produce a final product for end users.

25. The closest commercial precursor to the Internet existed in the Bulletin Board industry, which generated several hundred million dollars of revenue before the commercial Internet blossomed and replaced it. The sentence used 1989 as a marker because this is the year of entry for the first carriers for Internet traffic, PSINET and UUNET.

26. See Goldfarb (2004). This discussion follows norms at the U.S. Census, as expressed in the *Annual Service Survey.* Most households devoted most of its Internet budget to access fees (largely for services provided by Wi-Fi hot spots, or dial-up, broadband, wireless carriers) as opposed to subscription fees for content (largely provided by services such as Lexis-Nexis, the *New York Times* archive or *Wall Street Journal* archive). AOL sought to blur the distinction between access and content with a "walled garden" strategy and successfully did so for a few years with its dial-up service. Later, it reduced the importance of its access fees, relying on advertising for most of its revenue. This distinction does not count electronic commerce revenue, namely, use of electronic channels to support purchase of a good (e.g., clothing) or what had been a nondigital good (e.g., music).

Table 6.2   **Internet access at U.S. households**

	1997	1998	1999	2000	2001	2002	2003	2004	2005	2006
Households[a]	103.0	104.0	105.0	106.0	107.0	108.0	109.0	110.0	111.0	112.0
Internet adopters[a]	19.1	27.2	35.5	44.0	53.8	56.7	59.5	66.0	73.3	81.8
Broadband adopters[a]	n.a.	n.a.	0.9	3.2	9.6	13.0	18.5	27.5	41.1	47.0
Dial-up adopters[a]	19.1	27.2	34.5	40.8	44.2	43.7	41.0	38.5	32.2	34.7
% adopters[b]	18.6	26.2	33.8	41.5	50.2	52.5	54.6	60.0	66.0	73.1

Source: Greenstein and McDevitt (2009).
Note: n.a. = not available.
[a]Millions of households.
[b]Percentage of total households.

in Internet Publishing and Broadcasting revenue, of which $2.9 is advertising revenue.[27] Advertising revenue is now growing at a more rapid pace than subscription and access fees, and it may exceed access revenue soon, but not as of this writing.

Widely dispersed market decisions lie behind this revenue growth, shown by the diffusion of Internet access to U.S. households (table 6.2).[28] Starting with fewer than 20 percent of households in 1997, the Internet diffused to more than 73.1 percent of households by 2006. Similar results obtain for the diffusion of the Internet to business.[29]

Straightforward economic factors determined these trends: dial-up became available first and diffused to more than half of U.S. households. Broadband emerged later as a higher quality and more expensive alternative, albeit one available in only a few places and from a limited set of providers. Over time, however, broadband became more reliable and more widely available, and as that happened, many households paid to upgrade their Internet service.

6.2.1   The Initial Wave of Value Creation

A closer examination of the historical record shows that this market arose in distinct waves of entry and exit—the first wave of entry occurred after the NSF opened the Internet to commercial users, coupled with the invention of the World Wide Web and the creation of the World Wide Web Consortium,

27. See table 3.0.1, Information Sector (NAICS 51) or 3.4.2, Web Search Portals (NAICS 518112) in the 2007 *Service Annual Survey,* NAICS 51, Information, http://www.census .gov/svsd/www/services/sas/sas_data/sas51.htm, downloaded September, 2009. See table 3.3.5, Internet Publishing and Broadcasting (NAICS 516) in the 2007 *Service Annual Survey,* NAICS 51, Information. There is negligible adverting listed for Internet Service Providers (other than cable, telephone or wireless carriers), table 3.4.1. See http://www.census .gov/svsd/www/services/sas/sas_data/sas51.htm, downloaded September, 2009.

28. These data sources are described in more detail in Greenstein and McDevitt (2009).

29. As measured by the Current Population Survey (CPS) supplement and the Pew Survey of the Internet and American Life. See Forman, Goldfarb, and Greenstein (2003a,b).

which came into creation in 1994 to 1995. The World Wide Web, invented by Tim Berners-Lee, is an Internet application that links documents together through the use of hypertext and viewed by a Web browser. The commercial browser began to diffuse in 1995, enabling new functionality and new businesses built around this new technology. The Web quickly became the software platform for many creations thereafter, motivating further experimentation and magnifying the potential for value creation from the first wave of entrants.

The creation of the commercial browser caused a change of expectations about what was possible to do on the newly privatized Internet. The browser began as an academic project, but even that was sufficient to demonstrate an entirely new range of applications affiliated with linking various pages, displaying multimedia, and supporting a whole new interface for human-computing interactions.

Participants expected an explosion of commercial activity by established firms, venture capitalists, Wall Street analysts, and entrepreneurs, and, indeed, an immense entrepreneurial response did occur, which extended across a broad array of activities and applications, media, travel, commercial transactions, communications, and so on. The wave was a market response to new opportunity. Many different market participants sought to figure out how to apply the new technology to improve services to users. Indeed, unrestricted and entrepreneurial markets applied and reapplied these technologies over and over again to a wide range of problems and new applications.

During the wave of entry, new knowledge and lessons were shared at low cost.[30] Several distinct models emerged taking advantage of the demand for electronic services. One prominent model subsidized the delivery of text and other visual media with advertising. Many of the adherents to online news, entertainment, and other information-based commerce found this to be attractive. Another prominent model used the Internet for the delivery of a service, such as the creation of online retailers like Amazon and the addition of an online counterpart to other branded catalog retailers. Other models included developing a subscription service (such as for the *New York Times* crossword puzzle); organizing a place for buyer and seller to conduct a transaction, such as an auction, and charging a fee for the service; organizing a fee-based listing service, such as an online help-wanted listing; providing a fee-based matching service, such as for singles; and providing a location for aggregating information from users (e.g., blogs, recommendations, wikis), supported by advertising.[31]

---

30. The primary cost to society were the "co-invention costs," that is, the expenses incurred by suppliers and buyers in the pursuit of customizing the general purpose technology to the unique needs and idiosyncratic circumstances their market participants faced. These costs arose for users trying to apply the technology and suppliers trying to sell it. For a discussion of co-invention, see Bresnahan and Greenstein (1996).

31. For a summary of the diversity of models, see Hanson (2008) and Kirsch and Goldfarb (2008).

The first generation of browser and Internet Service Providers diffused extraordinarily rapidly. For example, the fraction of U.S. households online jumped from 18.6 percent in 1997 to 41.5 just three years later (table 6.2). The fast uptake of several popular applications of the 1990s (e.g., Hotmail, ICQ, and Yahoo!) reinforced this rapid diffusion.

However, the late 1990s saw more entry than actual demand would support a few years later. And so a shake-out ensued, first affecting access providers in 2000 (popularly known as the "Telecom Meltdown") and then eventually many online retailers (popularly known as the "dot-com crash"). Investment uncertainty after the events of September 11, 2001, magnified the downturn affiliated with this adjustment. Sellers with high debt and low revenue exited. This occurred at all levels of the value chain for Internet services, as well as infrastructure building.

While the mass exit led to widespread losses for many entrepreneurs and investors in entrepreneurs, with the benefit of hindsight the pattern of boom and bust should not come as a surprise. Much of the activity was exploratory in nature, and, by design, some explorations fail while others succeed. Moreover, historians of technology had described investment booms and busts for other episodes of technological innovations and commercialization, such as followed the growth of the railroads, the growth of the steel industry, the growth of the automobile industry, the growth of electrical power, and so on. Finally, computing markets also had experienced boom and busts during the development and deployment of the personal computer, the minicomputer, and client-server systems, albeit at the smaller scale than what followed the commercialization of the Internet.

The drama of the decline obscured another trend, how the first wave of experiments in value creation left a changed economic landscape. A large array of online activities survived, including large providers (e.g., AOL, Yahoo!, eBay, Google), as well as a wide array of niche products and services and productivity enhancements. Many catalog retailers successfully transitioned into online retailers, such as Victoria's Secret and L.L. Bean, and thrived just fine with their existing brand names and efficient order fulfillment. In short, even with excess entry, markets have a way of rapidly creating thriving businesses that take advantage of the opportunities enabled by the new technology. The results are hard to foresee until supply, demand, and prices plays itself out in all its glorious unpredictability.

6.2.2   Accumulating Innovation in the Internet

Accumulation of innovation in a market setting differs substantially from that in a skunk works.[32] In a market setting, there are common signs of healthy innovative behavior, even in a quickly evolving industry such as the Internet, and these underpin value creation by many participants. Commer-

---

32. Much of this section and the next provide a synopsis of arguments in Greenstein (2010b).

cial behavior resides inside a complex value chain. No single firm controls the value chain. The quality, price, and user experience arise from the interactions between participants in the value chain.

Even when there is no agreement about which criteria observers should use to assess the performance of the commercial Internet, there are patterns of healthy conduct, that is, commercial behavior indicative of an innovative industry. Such healthy behavior correlates with desirable marketwide outcomes, such as improvement in products, lower prices, new capabilities, or other innovations that lead to productivity improvements among business users.

Three general features of the market foster accumulation of innovation from value creating activities. These are economic experimentation, entrepreneurial initiative, and vigorous standards competition. *Economic experimentation* is a market-oriented action designed to help a firm learn or resolve uncertainty about an unknown economic factor. Usually such lessons cannot be learned in a laboratory or controlled environment, either because they involve learning about the nuances of market demand or learning about sets of procedures for providing new services at a lower cost.[33]

Not all economic experiments come with the same orientation or learning goal. Some focus on learning about the profitability of incremental changes in business processes. Some seek to learn about the restructuring of organizations and the profitability that may result from the simultaneous alteration of many processes. Some even seek to learn about the profitability of restructuring the relationship among many organizations within an industry.

Internet markets have been full of economic experiments in the last fifteen years. That was especially so in the latter part of the 1990s, when firms took a wide variety of bets to learn about unknown aspects of customer demand and the costs for meeting them using Web technologies, such as the browser, server software to support it, and a range of other innovations. These experiments covered all parts of the value chain for delivering services—Internet access, client-server platforms, contracting among business partners, and so on. Carriers conducted them, and so did content providers.[34]

Entrepreneurial initiatives involve an organization in a risky and challenging business in pursuit of a new economic opportunity. These firms are the market "participants" that make the first brave attempts at deploying, distributing, or servicing a new good to a wide range of customers. Small start-ups take entrepreneurial action and so do large firms. Sometimes small businesses that take such risks are bought by large organizations. Sometimes small start-ups go public and grow into large firms themselves. The increas-

---

33. *Economic experiments* pertain to any market experience that alters knowledge about the market value of a good or service (Rosenberg 1994; Stern 2006). Firms engage in economic experiments to reduce uncertainties about market value.

34. See Greenstein (2008a) for an examination of the role of economic experiments in the evolution of Internet access.

ing presence of entrepreneurs in communications markets has brought rapid change to many submarkets.

Yet entrepreneurial activity can increase and decrease for distinct reasons. Experimentation may lead to entrepreneurial initiatives by entrants, or it may enhance the products of one particular firm. It forces incumbents to react, or, even better yet, anticipate the entrant and innovate in advance. This fosters incentives to lower prices and sponsor more innovative products, and sooner. Users benefit from all of those.

Vigorous standards competition also played a role in innovation in the Internet. That is because leading-edge technologies often cannot deploy on a wide scale without some routines or processes or coordination of activities across many firms. Thus, the ratification of new standards generally acts as a leading indicator of impending technological progress and serves as another sign of a healthy innovative industry. While new standards and upgrades to existing standards may not arrive at a regular rate, a slow pace for development or a slow arrival of new standards usually sets off alarms.

To be sure, this benchmark is particularly challenging to put into practice because some standards are more important than others. The Internet Protocol Suite known as TCP/IP have played a central role for decades, for example, and any alteration to them receives considerable attention, deservedly more attention than other standards. The same is true for protocols that govern the World Wide Web, which are handled at the World Wide Web Consortium (W3C). This is also so for important components of the Internet, such as upgrades to wire-line Ethernet. That topic is discussed at the Institute of Electrical and Electronics Engineers (IEEE) Standards Association committees assigned to new standards. In the case of wire-line Ethernet, for example, it tends to be subcommittees of the Working Group for Wireless Local Area Network Standards.

Standards design needs competition. Although the process of standards design in which market competition has played a role can be a messy, frustrating, and confusing process, this mess is necessary. Standards designed in the absence of competition usually have been orderly, infrequent, and simplified. Such standards have been more likely to lead an industry down as unhealthy an innovative path as it can go.

If a firm with market power designs a new standard, it will face strong incentives to roll it out slowly to protect the firm from cannibalizing its own monopoly rents. For example, in the days when IBM controlled a large part of the mainframe market, it could not bring itself to abandon Extended Binary Coded Decimal Interchange Code (EBCDIC), its standardized proprietary language, or, for that matter, to help others migrate up from EBCDIC to the many other superior languages available. Despite plenty of improvements IBM could have made, its managers refused to deploy them, preferring instead to exploit locked-in users.[35]

---

35. See Brock (1975).

Monopolies also face strong incentives to have a "quiet life," to para-phrase Sir John Hicks.[36] That is, monopolies may exert less effort when they choose standards or design them to castrate user choices in such a way that leads to less inconvenience for the monopolist at the expense of the user (e.g., trimming product line breadth or trimming away complex attributes of the product). For example, until the mid-1970s, AT&T held a monopoly over residential customer telephone handsets. Most households faced a lim-ited menu of (overengineered and excessively rigid) choices. Well-engineered or not, there were too few choices in comparison with what a competitive market would have done.

With the breakup of AT&T's monopoly, multiple providers began to match the offerings of its nearest rivals. In a short time, the heated and urgent competitive behavior familiar to consumer electronics eventually overtook the market, leading to a plethora of choices at a wide range of prices. In other words, in the absence of restraining limitations on discre-tion, monopolies have designed selfish standards. An antidote to the selfish standards of monopolies has been competition between standards. In the history of the Internet, massive entrepreneurial entry drove innovation, and accessible standards contributed to it.

6.2.3   Negotiations between Open and Proprietary Standards

One feature of the competitive Internet is probably the most crucial for accumulating innovation from dispersed market participants. Not surpris-ingly, it is the most controversial. *Negotiations over interdependent processes* shaped how the market accumulated services and built on each other. These negotiations took on importance because every participant, the innovative and not so innovative, operated within a system of technically interrelated components and services where these processes interoperated. The failure or reduction in performance of any of these activities could lead to degradation of the quality of outcomes for many users.

In a network with a high degree of technical interrelatedness, there are general gains to all parties from bringing routines into business processes and activities, much like there are gains to adopting standards and platforms to coordinate activities. While there may be no better way to reduce complex-ity, adopting such routines may require negotiation between multiple parties. For example, even the simplest of activities, such as sending e-mail, involves many participants, and efficient delivery of services depends on advanced agreement about how their business activities will interrelate.

To reduce the uncertainty about how such services interoperate, commer-cial firms take one of two approaches, either they negotiate arrangements

---

36. "People in monopolistic positions . . . are likely to exploit their advantage much more by not bothering to get very near the position of maximum profit, than by straining themselves to get very close to it. The best of all *monopoly* profits is a *quiet life*" Hicks (1935, 8).

(contractual norms) in advance with all relevant participants, or they do it all themselves by offering a platform (a bundle of standards) that accomplishes the same task, internalizing the contracts within one firm's decision making.

Although the inception of the early Internet was a "network of networks," today leading firms and their business partners view the commercial Internet as a "network of platforms." This seemingly small change in definition is far from insignificant. The rise of a plethora of platforms on the Internet is a source of both celebration and consternation. Platforms perform functions that firms or users value. Their presence usually suggests that some firms or users are better off with them than without, and it usually suggests they have replaced an inadequate nonproprietary standard inherited from the era prior to commercialization. At the same time, large or dominant platform leaders usually possess market power, and that occasionally gives them the ability to resist nonproprietary standards that serve the interests of some rivals.

Which is better, proprietary or open? Such debates inevitably boil down to restrictions on the discretion of incumbent management to determine standards. Proprietary and open standards contrast most sharply in their respective approaches to transparency and participation.

With standard proprietary platforms, leading firms retained discretion and guided participation within strict rules. Generally, strong platform leaders, such as IBM, Microsoft, and Intel, retained their authority by owning assets on which others depended and by not being transparent about how such assets would change in the near future. Such practices came into direct conflict with the transparent and participatory processes for standards development in the Internet, particularly as practiced at the Internet Engineering Task Force (IETF), which used these processes to support group decision making.[37]

Conversely, transparent processes are those in which decision makers alert participants to imminent change—sometimes well in advance—when their change will diminish the returns on others' innovative investments. In many Internet standards forums, such as the IETF, the organizations take considerable effort to remain transparent and embed such norms in the operations of the group.[38] Such transparency is one of the reasons why standards processes have become a strong indicator of the imminent release of leading-edge technologies in Internet equipment. Interested parties monitor the designs (because they can attend IETF meetings) and know that their near rivals do the same (because the data are available to anyone). All parties plan to match each other along the dimension of the standard and

37. These trade-offs are discussed at length in Greenstein (2009), which contrasts the operations of the Microsoft, as it developed Windows 95, with the operations of the IETF, as it operated when the privatization of the Internet first began.
38. See Bradner (1999).

differentiate along the dimensions in which each has competitive advantage. Competition ensues once the standard is upgraded from its Beta to an endorsed and official standard.

Transparency is also a feature found frequently in open source projects with importance to the Internet value chain, such as Linux, Apache, Firefox, and the W3C. In the experience of Internet networks, a minimal level of transparency has been a necessary element of an open value chain operating for a large number of users. Defining such minimal levels is an important economic issue if other firms will not make long-term investments unless they understand at a fine level of detail how their software must interact with another's.

Transparency is distinct from participatory rules.[39] Participatory processes are those in which sponsoring organizations invite comment, discussion, and input from others affected by their actions. Such organizations solicit input through public forums, e-mail lists, blogs, community sites, and a range of other activities. Standards organizations vary considerably in their policies for encouraging or discouraging participation. Some organizations charge fees, some require participants to meet certain technical qualifications, and others allow any observer to attend but not vote. For example, the organization that designed and updates Wi-Fi does not allow unrestricted participation; firms must pay a fee in order to send representatives.

Wide participation is found quite frequently in open source projects, particularly those without sponsorship. Often technical skill determines participation. For example, the Firefox browser community has quite diverse participation from numerous corners, though participants tend to self-select on the basis of technical skill simply because they would be lost otherwise. Similar observations hold for Linux and Apache. In both examples, most participants are quite technically skilled, and in the latter case, such skill acts as an explicit qualification. Wikipedia is perhaps the best-known example of an online project that encourages wide participation from a community of contributors and where no skill test is applied to contributors.

Wide participation is probably the least common attribute among standards consortia sponsored by commercial firms. Most managers prefer to retain decision-making authority, guarding investment decisions in the name of stockholders. There is concern that giving up such discretion risks having participants take investment in directions that do not serve firm interests.[40] In addition, accommodating wide participation normally comes at a cost, such as slower decision making and more onerous managerial challenges in coming to consensus. Hence, even some ostensibly open standards processes chose to restrict participation. For example, Tim Berners-Lee established the

---

39. See West and O'Mahoney (2008).
40. See West and O'Mahoney (2008).

W3C with a less participatory structure than found in the IETF, where he had personally experienced the drawbacks of slow decision making when he first tried to standardize the core inventions behind the World Wide Web.[41]

Although the Internet experience does not give precise directions toward the best choice for participation rules, in the past, wider participation has tended to beat out no participation. Thus, every proprietary platform adopts some degree and form of participation (though, to be sure, with varying degrees of transparency). Perhaps the biggest surprise from the Internet experience is the persistence of standards-making institutions with wide participation and transparent processes. Once established, these institutions have persisted, coexisting alongside proprietary platforms, sometimes as competitors and sometimes as complements. Firms have learned to live with these institutions, and many firms have learned to thrive alongside them. Cisco, Intel, IBM, and many Wi-Fi firms are active participants in these standards forums. Even (previously reluctant) firms such as Microsoft, AT&T, and Verizon have found it useful to participate and fund such activities.

These institutions guided the accumulation of innovations in a market setting. Successful platforms accumulate additional functionality over time. Leaders of platforms with proprietary interests attempt to grow the functionality for their platform, as well as direct the gains from growth to their own firms. Information sharing and flow between participants is an instrument in achieving those goals. In contrast, in nonproprietary settings, the accumulation of innovation differs because the information accumulates in organizations that foster transparency and, often, do not place restrictions on the use of the information. During the initial growth of the commercial Internet, both types of organizational forms—as sponsored by Microsoft in Windows 95 or the World Wide Web Consortium, for example—had successful innovative experiences. Both nurtured big innovative pushes, accumulation of incremental innovation from multiple sources, and impressive value creation after coordinating innovation from multiple sources.[42]

6.3 Policy and Governance

Direct government support for R&D in the creation of the Internet had two potential effects. It accelerated the arrival of the technology, and it influenced its direction. During the precommercial era of the Internet, the creation of a general purpose technology for exchanging packets of data between many firms was risky. It had no immediate obvious commercial payoff. Program officers at DARPA intended to fund radical technological progress that otherwise would not have been funded by private firms. They

---

41. This frustration is described in detail in Berners-Lee and Fischetti (1999).
42. See Greenstein (2009).

intended to develop research communities in those areas where almost none had existed. The investment aimed to develop fundamental scientific understanding and engineering experience, accelerating the arrival of actual products and services at some point. The Department of Advanced Research Projects Agency accomplished its goals and then some. Long before the Internet arrived in particular, packet switching was but a theoretical idea and expensive to implement. There is no doubt the initial work funded by DARPA in the 1970s accelerated the arrival of the technology. No other private firm at the time, such as IBM or AT&T, had projects in the area coming close to DARPA's efforts in size and scope.[43]

The Department of Advanced Research Projects Agency also funded the building of research prototypes and the building of a prototypical system. Arguably, that went beyond the aspirations of the DOD but was an immediate by-product of the project's success and a natural extension in terms of extending the scientific/engineering frontier. It had enormous value, too, demonstrating the feasibility of what had been a theoretical idea.

In an industry where many potential future technologies vie for attention, there was value in demonstrating that one of these forecasts was viable. In this case, DARPAnet, and later NSFNET, showed that the successful operation of the entire system could have great value. This illustrates how demonstrations can serve as a focal point for further development, particularly in the face of widespread industry resistance prior to such demonstration.

However, government funding came with a drawback. Restrictions on participants and "acceptable use" truncated experimentation and entrepreneurial initiatives. This truncation was a detriment to understanding the potential for the network outside of the limited uses to which DARPA's community put it.

The government's role differed in the later time period, particularly the era just prior to the blossoming of the commercial Internet. In the private sector by the late 1980s, most savvy commercial observers anticipated the arrival of mass market electronic commerce, but few anticipated that the Internet would be it, and, for related reasons, few guessed that any would arrive as soon as did the Internet.

Moreover, many private firms largely ignored the Internet. That fact colors any interpretation of the government's role. Why did many private firms ignore investing heavily in the Internet? The skepticism at many corporations can be interpreted in three (overlapping) ways:

- A misunderstanding of the potential for the Internet, perhaps due to the commitment to an alternative technological vision or forecast.
- A situation in which the Internet lacked internal "champions" inside leading organizations, perhaps because of the expectation that it would cannibalize too many revenue streams at existing business.

43. See Abbate (1999).

- A situation in which the Internet benefited many users at once, perhaps because no single firm had incentive to nurture adoption that seemingly did not directly contribute to their own bottom line.

For example, neither AT&T's management nor that at any of the "baby Bells" expressed any strategic interest in commercializing services related to TCP/IP in the late 1980s. Most preferred to invest in services such as the Integrated Services Digital Network (ISDN), which the managers considered the technical direction worth exploring.[44]

To be fair to AT&T's managers, they were not alone. For example, despite employing some of the best researchers in the world on this topic and despite the involvement of its research division in the NSFNET, IBM's strategic planning at the corporate level also ignored the Internet. The company did not aggressively commercialize related services. Instead, its corporate plans called for commercializing a proprietary set of networking technologies, built around its Systems Network Architecture (SNA). These plans ultimately led to some of the most high-profile product development failures in IBM's history.[45] Likewise, Digital Equipment Corporation, then the second largest computer company in the world, was strongly committed to DECNet, a proprietary network service.

Even many sponsors of the NSFNET ignored its commercial potential. Most of the carriers were holdovers from the NSF, such MCI and Sprint and BBN's division. Only a few other participants from the NSF era took entrepreneurial actions aimed at the commercial market, such as the managers who started carriers PSINet and UUNet, who entered in 1989, far before many others. Only a few bulletin board providers, such as Prodigy and CompuServe, made early switches to the Internet.[46] For example, none of the other midlevel carriers made a switch to for-profit status until *after* the NSF commercialization neared completion in 1994—five years later. As it turned out, AT&T did not start to offer Internet service until 1995, which is about the same time as many other mainstream firms. First it offered service to business and then to homes a year later. It continued to do well with business users as well as briefly with home users before it faded later in competition with AOL. The midlevel networks started to convert to for-profit status about the same time. Most of the baby Bells were even later than that.

6.3.1   Shaping Direction

The rise of the Internet shaped the direction of technical change. The changes contained two attributes: first, the network was comprised largely of nonproprietary features/protocols/standards coupled with an open organ-

44. ISDN is a circuit-switched telephone network system that can carry voice as well as data over the same line.
45. See Gerstner (2002).
46. Though, for a number of reasons, it largely did not help them gain market share or thrive during the first wave of entry. This story is told particularly well by Banks (2008).

ization to support existing standards and update them (e.g., the IETF and W3C). Its existence was unexpected, in part because its leadership structure differed from every other alternative considered plausible by contemporary executives—such as a network dominated by any established firm (e.g., IBM), carrier (e.g., AT&T), equipment manufacturer (e.g., Lucent), quasi-government agency (e.g., the ITU), or industry consortium (e.g., ISO).

For example, although many firms had e-mail services for their own computer networks, none of them had incentives to combine their systems with others'. No single firm had incentives to aggregate innovative suggestions from a vast array of contributors at the early and risky stage of developments. The Internet's nonproprietary features acted as an attribute around which many participants could agree because none of them individually risked too much nor benefited too much.

Coupled with the open and nonproprietary nature of the Internet was a surprising set of technological leaders. Although government support made no difference to the stature of early innovators like Paul Baran, Joseph Licklider, or Leonard Kleinrock, whose reputations would have been high with or without DARPA's projects, that was not true for the first generation of Internet developers, such as Steve Crocker or Vint Cerf. They often expressed surprise at the discretion they had, many of them becoming leaders as graduate students. Their historical recollections refer to many moments when they wondered when they would be displaced by "a professional crew," that is, more senior researchers in the field of computer science.[47]

Perhaps the biggest change in direction came from the structure of governance that came along with the Internet. The IETF and W3C, among others, were open processes, in the sense that they fully documented their activities, did not restrict participation, and never actively sought to exclude any innovator from building applications on to the installed base of accumulated protocols.

Not all aspects of that open structure were important.[48] For example, plenty of industry experience suggested that commercial organizations producing proprietary hardware and software designs could be as innovative as open communities. Sometimes open communities have been more innovative, and sometimes proprietary firms have been, and both have coexisted. Rather, open institutions had two key structural features: not withholding information and not restricting its use. These features enabled the World Wide Web to commercialize so quickly. More pointedly, the IETF's leadership was unwilling to withhold information from anyone, effectively not excluding outsiders, such as Tim Berners-Lee at the time, even though his inventions potentially displaced so many established processes and existing technologies, even technologies supported by the IETF.[49] Once the World

47. See Crocker (1987).
48. This argument is developed more fully in Greenstein (2009).
49. Even though the vast majority of participants inside the IETF viewed Tim Berners-Lee's proposal with indifference or hostility, all were perfectly willing to let him use all of the IETF's

Wide Web began to diffuse, no established firm could stop others from building on it and bringing about a massive change in many aspects of economic activity.

6.3.2   Governance of the Rules of the Game

Experience in this industry highlights the importance of good governance—simply spending federal R&D money or adopting policies from a check list, by itself, would not have been sufficient to achieve success. Rather, successful public support for innovation has been embedded in an institutional structure that provided checks and balances, counterbalancing the risk of any effort from degenerating into pork barrel spending and into coddling of existing incumbents. Creating this kind of system has required time, judgment, and (sometimes) strong political will.

During the Internet's precommercial era, many issues required sound judgments by public servants who were focused on executing a vision of what they thought would benefit the technological development of the Internet. Indeed, a crucial feature of DARPA's success resided in stating a clear mission for its effort.[50] Another involved choosing managers with extraordinary intelligence and competence, giving them funds and discretion, and allowing them to work with minimal oversight.[51] Managers played a crucial role at both DARPA and the NSF, but they did not act alone. They had support from their direct supervisors and their coworkers, all of whom could articulate their general mission and understand how that translated into short purposeful managerial action. The precommercial era of the Internet also received political support from those in the defense department committed to DARPA's autonomy.[52] Political actors did not intervene in the research involved in the Internet, although the Mansfield Amendment did influence a number of other related projects funded by DARPA. Political management also supported the NSF's stewardship and beyond (e.g., support from Senator Al Gore and Congressman Rick Boucher).[53]

During the commercial era, government played a role in setting the "rules of the game" by shaping negotiations among participants. In particular, le-

---

tools. Berners-Lee, thus, built on top of existing IETF approved protocols with full freedom and discretion. Thus, he was able to take action quickly. See Berners-Lee and Fischetti (1999).

50. Licklider's three criteria for funding research still sound prescient today: "1. The research must be excellent research as evaluated from a scientific or technical point of view; 2. The research must offer a good prospect of solving problems that are of interest to the Department of Defense; 3. The various sponsored efforts must fit together into one or more coherent programs that will provide a mechanism, not only for execution of the research, but also for bringing to bear upon the operations in the Defense Department the applicable results of the research *and* knowledge and methods that have been developed in the fields in which the research is carried out" Norberg, O'Neill, and Freedman (1996, 29). See also Waldrop (2001) for a wider discussion.

51. See Norberg, O'Neill, and Freedman (1996).

52. See Norberg, O'Neill, and Freedman (1996).

53. Wiggins (2000) provides an overview of Al Gore's role in securing funding for the NSF, and Segaller (1998) partially recounts Boucher's role in opening the Internet to commercial use. For the latter, see also Shah and Kesan (2001).

gal questions covering intellectual property, monopoly powers, and other limitations and protections have shaped the Internet landscape. Market actors are sensitive to persistent and unresolved legal uncertainty over liability, ownership, and other legal rules that shape returns on investment. Hence, crucial parts of the value chain for the Internet have stalled as participants awaited legal or regulatory rulings settling boundaries.[54] Recently, for example, YouTube was founded in an era when there were multiple plausible definitions for the precise legal safe harbor for including copyrighted material on a Web site for user-supplied video. These definitions today still remain ambiguous. Google acquired YouTube in spite of the shadow of the legal risk, and its investments (worth hundreds of millions of dollars) will most likely change as court decisions change.

After the retirement of the NSFNET and during the massive investments in the commercial Internet, it was fashionable to claim that the government's role was minimal in fostering innovative incentives. Such a claim is fatuous at best. Government actors were involved in determining rules for and resolving disputes about the minimal technical requirements telephone companies had to follow when interconnecting with a dial-up Internet Service Provider, and these were crucial for fostering the development of the early industry. Government actors also required a divestiture of assets as a condition for merger when WorldCom sought to merge with MCI, thus thwarting aspirations to assemble a large fraction of the Internet's backbone under one organization. And, perhaps better known, government actors were involved in a wide array of issues that arose around Microsoft's behavior during the browser wars.

Such debates quickly reached back into the institutions that governed standards for the Internet. For example, legal precedents were set at a wide array of government organizations with jurisdiction over these disputes, such as at the Department of Justice, Federal Trade Commission, Federal Communications Commission in the United States, and at equivalent European Union regulatory bodies.

The policies that have resulted from legal battles have been a source of regulatory tension and friction. For example, established regulations, known as Computer II, compelled the U.S. phone industry to accommodate the new Internet Service Provider (ISP) industry. Managers in the existing telephone firms did not want to accommodate dial-up ISPs, but did so at first because Computer II required it of them.

But policies such as Computer II were not there to support the Internet. Rather, they were the outgrowth of two long-standing principles: (a) common carrier regulation for telephones, which prevented the telephone company from being selective about who they served; and (b) antitrust regu-

---

54. See the analysis of such matters in the area of communications carriers in, for example, Goldstein (2005).

lations, which had led to the divestiture of AT&T, and, more important, to a series of regulations for governing carrier interactions with others, such as equipment firms and providers of services over telephone lines, such as Bulletin Board providers.[55] Because of these legal actions, the United States had a less hostile approach to entrepreneurial entry of dial-up ISPs than did most of the world for reasons unrelated to the Internet in particular.

The United States also had a very nurturing legal regime for consortia and standard-setting bodies. At crucial moments, these policies fostered a healthy dose of vigorous standards competition. Once again, these policies existed for their own reasons and not because any policymaker was trying to encourage the Internet in particular. For example, such laws played a crucial role in Tim Berners-Lee's personal decision making. In the United States, he received a much more welcoming set of conditions than those he faced in Switzerland, motivating him to leave employment at the European Organization for Nuclear Research (CERN) and establish the W3C at Massachusetts Institute of Technology (MIT).

6.4   Finally, Why the Internet Worked

The history of the Internet highlights two distinct ways of organizing a long-term program for accumulating innovation in a complex interdependent system. One approach relies on autonomous research institutions (skunk works) to organize and nurture inventive employees (wild ducks). The other approach relies on commercial markets to aggregate dispersed initiatives from a wide array of entrepreneurial participants.

A skunk works faces a significant danger of innovating into areas where there is no demand, and, thus, no economic value. How did the precommercial Internet create value in spite of the absence of commercial demand? First, it avoided some dangers by keeping prototype and operations sufficiently close to one another. The first participants in the noncommercial Internet assessed value from their own experiences, and DARPA managers nurtured and permitted experimentation to blossom. That helped them create useful and innovative applications such as e-mail and packet switching. The DARPA skunk works worked within community norms that fostered accumulating technologies on the merits, avoiding technical dead ends. In addition, DARPA and the NSF played a pivotal role in becoming the element of "demand" for which innovation was supplied. The agencies' substantial funding to research institutions (as well as leveraged funds through distributed investments to universities) procured the innovation for what became the breakthrough technologies leading to today's Internet. This investment was not easy, automatic, or inexpensive, but many would

55. For a summary of the consequences of these rules for dial-up ISPs in the United States, see Goldstein (2005) or Greenstein (2008b).

argue it has been one of the most important innovative undertakings supported by the U.S. government.

On the other hand, the skunk works approach restricted participation and truncated experimentation by excluding innovation along lines that did not support the "acceptable use" requirements of the government agencies. Such restrictions limited learning to an artificially narrow range of issues and left a wide array of other applications untouched.

In contrast, the commercial era of the Internet played to the strength of market-based innovation. It permitted decentralized exploration from commercial firms facing a wide array of incentives and a wide variety of idiosyncratic circumstances. Market-oriented exploration did a marvelous job of exploring the range of uncertain factors affiliated with satisfying demand, thus demonstrating the benefits of conducting many economic experiments. Once released to commercial interests, the Internet became the springboard for a dizzying array of applications that were not envisioned by the sponsoring government agencies. These applications, particularly the World Wide Web and its associated browsing technology, quickly infiltrated nearly every aspect of U.S. business and domestic life, and their effect continues to grow.

However, these explorations came to fruition because they were built upon a backbone technology that no single player or group of players in the market was willing or able to undertake—or, for that matter, were forward-thinking enough to even visualize it. Throughout the history of the Internet, standards, protocols, and other rules of governance have shaped the direction and rate of innovations emerging from it. Some of these guiding factors grew with the project, most notably standards such as TCP/IP and protocols that govern the World Wide Web, as well as standard-setting bodies such as the W3C and IEEE. Other influential forces were not specific to the Internet but shaped it markedly, such as Computer II and legal rulings against monopoly control over communications technologies.

Perhaps because the DARPA skunk works invested heavily in many different directions (and many of them ultimately not bearing fruit) and used the brain power of many researchers at an array of institutions, they themselves garnered the power of decentralization (which is usually affiliated with a marketplace), albeit in narrower, more-disciplined form. Similarly, because the commercial Internet relied on strict protocols and standards-setting bodies, they, in effect, demonstrated the discipline seen more typically in centrally funded efforts.

While the government-based approach to innovation and the market-oriented approach each have their strengths and challenges, in the case of the Internet, these two systems came together in a unique, phased, and ultimately complementary way. The accumulated knowledge enabled the creation of value in myriad numbers of applications that continue to shape the world around us all.

References

Abbate, Janet. 1999. *Inventing the Internet.* Cambridge, MA: MIT Press.

Allen, Robert C. 1983. Collective invention. *Journal of Economic Behavior and Organizations* 4 (1): 1–24.

Banks, Michael A. 2008. *On the way to the Web: The secret history of the Internet and its founders.* Berkeley, CA: Apress.

Berners-Lee, Tim, with Mark Fischetti. 1999. *Weaving the Web, the original design and ultimate destiny of the World Wide Web by its inventor.* New York: Harper Collins.

Bradner, Scott. 1999. The Internet Engineering Task Force. In *Open sources: Voices from the open source revolution,* ed. Chris DiBona, Sam Ockman, and Mark Stone, 47–52. Sebastapol, CA, O'Reilly Media.

Bresnahan, Timothy, and Shane Greenstein. 1996. Technical progress and co-invention in computing and in the use of computers. *Brookings Papers on Economics Activity: Microeconomics:* 1–78.

Brock, Gerald. 1975. Competition, standards and self-regulation in the computer industry. In *Regulating the product: Quality and variety,* ed. Richard Caves and Marc Roberts, 75–96. Cambridge, MA: Ballinger.

Clark, David. 2008. Personal communication between David Clark and the author, September 26.

Crocker, David. 2008a. Personal communication between David Crocker and the author, August 7.

———. 2008b. A personal view: The impact of email work done at the RAND in the mid-1970s. Brandenburg InternetWorking. Mimeograph. http://www.bbiw.net/articles/rand-email.pdf.

Crocker, Steven D. 1987. The origins of RFCs. In *RFC 1000—Request for comments reference guide,* compiled by J. Reynolds and J. Postel. http://www.ietf.org/rfc/rfc1000.txt.

Forman, Chris, Avi Goldfarb, and Shane Greenstein. 2003a. The geographic dispersion of commercial Internet use. In *Rethinking rights and regulations: Institutional responses to new communication technologies,* ed. Steve Wildman and Lorrie Cranor, 113–45. Cambridge, MA: MIT Press.

———. 2003b. Which industries use the Internet? In *Organizing the new industrial economy,* ed. Michael Baye, 47–72. Amsterdam: Elsevier.

Frazier, Karen. 1995. *Building the NSFNet: A partnership in high speed networking.* Ann Arbor, MI: Merit, Inc. http://www.merit.edu/documents/pdf.nsfnet/nsfnet_report.pdf.

Gerstner, Louis V., Jr. 2002. *Who says elephants can't dance? Inside IBM's historic turnaround.* New York: HarperBusiness.

Goldfarb, Avi. 2004. Concentration in advertising supported on-line markets: An empirical approach. *Economics of Innovation and New Technology* 13 (6): 581–94.

Goldstein, Fred. 2005. *The great telecom meltdown.* Boston: Artech House.

Greenstein, Shane. 2008a. Economic experiments and industry know-how in Internet access markets. In *Innovation policy and the economy.* Vol. 8, ed. by Adam B. Jaffe, Josh Lerner, and Scott Stern, 59–110. Chicago: University of Chicago Press.

———. 2008b. Innovation and the evolution of market structure for Internet access in the United States. In *The Internet and American business,* ed. William Aspray and Paul E. Ceruzzi, 47–104. Cambridge, MA: MIT Press.

———. 2009. Open platform development and the commercial Internet. In *Plat-

forms, innovation and competition, ed. Annabelle Gawer, 219–50. Northampton, MA: Edward Elgar.

———. 2010a. The emergence of the Internet: Collective invention and wild ducks. *Industrial and Corporate Change,* doi:10.1093/icc/dtq047, http://icc.oxford journals.org/.

———. 2010b. Glimmers and signs of innovative health in the commercial Internet. *Journal of Telecommunication and High Technology Law* 8 (1): 25–78.

Greenstein, Shane, and Ryan McDevitt. 2009. The broadband bonus: Accounting for broadband's impact on U.S. GDP. NBER Working Paper no. 14758. Cambridge, MA: National Bureau of Economic Research.

Hanson, Ward. 2008. Discovering a role online: Brick-and-mortar retailers and the Internet. In *The Internet and American business,* ed. William Aspray and Paul E. Ceruzzi, 233–58. Cambridge, MA: MIT Press.

Hicks, John. 1935. Annual survey of economic theory. The theory of monopoly. *Econometrica* 3 (1): 1–20.

Kirsch, David, and Brent Goldfarb. 2008. Small idea, big ideas, bad ideas, good ideas: "Get big fast" and dot-com venture creation. In *The Internet and American business,* ed. William Aspray and Paul E. Ceruzzi, 259–76. Cambridge, MA: MIT Press.

Leiner, Barry, Vinton Cerf, David Clark, Robert Kahn, Leonard Kleinrock, Daniel Lynch, Jon Postel, Larry Roberts, and Stephen Wolff. 2003. *A brief history of the Internet, version 3.32.* Reston, VA: The Internet Society. http://www.isoc.org/internet/history/brief.shtml.

Mandelbaum, R., and P. A. Mandelbaum. 1992. The strategic future of mid-level networks. In *Building information infrastructure, issues in the development of the National Research and Education Network,* ed. Brian Kahin, 59–118. Cambridge, MA: McGraw-Hill Primis.

Maney, Kevin. 2003. *Thomas Watson Sr. and the making of IBM.* Hoboken, NJ: Wiley.

Meyer, Peter. 2003. Episodes of collective invention. U.S. Bureau of Labor Statistics Working Paper no. 368. Washington, DC: U.S. Bureau of Labor Statistics. http://ssrn.com/abstract=466880.

Norberg, Arthur, Judy O'Neill, and Kerry Freedman. 1996. *Transforming computer technology, information processing for the Pentagon, 1962–1986.* Baltimore, MD: Johns Hopkins University Press.

Partridge, Craig. 2008. The technical development of Internet email. *Annals of the History of Computing* 30 (2): 3–29.

Rich, Ben R., and Leo Janus. 1994. *Skunk works: A personal memoir of my years at Lockheed.* Boston, MA: Back Bay Books.

Roland, Alex, and Philip Shiman. 2002. *Strategic computing: DARPA and the quest for machine intelligence 1983–1993.* Cambridge, MA: MIT Press.

Rosenberg, Nathan. 1977. The direction of technological change: Inducement mechanisms and focusing devices. In *Perspectives on technology,* by Nathan Rosenberg, 108–25. Cambridge, UK: Cambridge University Press.

———. 1994. Economic Experiments. In *Exploring the black box: Technology, economics, and history,* by Nathan Rosenberg, 87–108. Cambridge, UK: Cambridge University Press.

Segaller, Stephen. 1998. *Nerds: A brief history of the Internet.* New York: TV Books.

Shah, Rajiv C., and Jay P. Kesan. 2001. Fool us once, shame on you—Fool us twice, shame on us: What we can learn from the privatizations of the Internet backbone network and the domain name system. *Washington University Law Quarterly* 79 (1): 89–220.

Stern, Scott. 2006. Economic experiments: The role of entrepreneurship in economic prosperity. *The Melbourne Review: A Journal of Business and Public Policy* 2 (2): 53–56.

Waldrop, Mitchell. 2001. *The dream machine: J. C. R. Licklider and the revolution that made computing personal.* New York: Penguin.

Watson, Thomas, Jr. 1963. *A business and its beliefs: The ideas that helped build IBM.* New York: McGraw Hill.

West, Joel, and Siobhan O'Mahoney. 2008. The role of participation architecture in growing open source communities. *Industry and Innovation* 15 (2): 145–68.

Wiggins, Richard. 2000. Al Gore and the creation of the Internet. *First Monday* 5 (10-2). http://www.firstmonday.org/issues/issue5_10/wiggins/.

Venture Capital and Innovation in Energy

Josh Lerner

7.1 Introduction

The past two years have seen challenging times for venture capital activity. The fact that no companies went public in the second quarter of 2008—the first time in three decades that this happened—and the low realized returns for venture funds in the past decade more generally have raised alarms about the viability of the venture model. As Dixon Doll, cofounder of Menlo Park-based DCM and current National Venture Capital Association chairman remarked:

> While we clearly recognize that the IPO drought is being driven largely by a weak economy, there are other systemic factors that are making the IPO exit less attractive for high quality venture-backed companies. Our government and the private sector should be doing all that it can to encourage these innovative, high quality companies to enter the public markets and grow from there.[1]

As a result of these questions, the volume of funding raised by venture capital organizations and the amount disbursed to portfolio firms have both dropped. In few places has this drop been as dramatic as in alternative

Josh Lerner is the Jacob H. Schiff Professor of Investment Banking at Harvard Business School, with a joint appointment in the Finance and the Entrepreneurial Management Units, and a research associate and codirector of the productivity program at the National Bureau of Economic Research.

I thank Harvard Business School's Division of Research for financial support. This essay is based in part on Gompers and Lerner (2001) and Lerner (2009). I thank Kathy Han for her invaluable assistance. Helpful comments were provided by participants in the Accelerating Innovation in Energy: Insights from Multiple Sectors conference, Rebecca Henderson, and an anonymous referee. All errors are my own.

1. See http://www.nvca.org/pdf/Q2_08_Exits_Release.pdf.

energy. The amount disbursed to these firms fell from \$1.3 billion in the second quarter of 2008 to just over \$200 million in the first quarter of 2009, and while there was some recovery by 2010, the investment levels remained considerably below earlier heights.[2]

Already voices have been raised, expressing worry about the implications of this decline—and the associated shifts in venture capitalist behavior—for technological innovation. For instance, in an influential new volume, Judy Estrin (2008), the former chief technology officer of Cisco systems, argues that short-term thinking and a reluctance to take risks are causing a noticeable lag in innovation. She argues that venture capitalists that back entrepreneurial firms have been too cautious to make big bets, particularly after the costly failures experienced during the dot-com bust.

This chapter seeks to understand the implications of the difficulties in the venture market on innovation, with particular emphasis on alternative energy. It makes three arguments:

- Venture capital funding has an important role to play in stimulating innovation and economic growth.
- But venture funding has a tendency to be cyclical. This tends to reduce the private and social returns to these innovations.
- These dynamics have important implications for thinking both about the probable effectiveness of private-sector investments in energy and whether and how the government should play a role.

In particular, the final section makes two key claims. First, it argues that the situation may not be as grim as it initially appears. While there are many reasons for believing that, on average, venture capital has a powerful impact on innovation, the impact is far from uniform. In particular, during boom periods, the prevalence of overfunding of particular sectors can lead to a sharp decline in terms of the effectiveness of venture funds. While prolonged downturns may eventually lead to good companies going unfunded, many of the dire predictions seem overstated.

Second, we consider some of the implications for public policy. Our analysis suggests that, while the rise of venture capital has been an important contributor to technological innovation and economic prosperity, an effective policy agenda going forward will not simply seek to spur much venture financing. We highlight the fact that many of the steps that policymakers have pursued have had the consequence of throwing "gasoline on the fire": that is, they have exacerbated the cyclical nature of venture funding. Instead, the environment for venture capital investment can be substantially improved by government policies (both federal and state) that encourage private investment and address "gaps" in the private funding

2. See https://www.pwcmoneytree.com/MTPublic/ns/moneytree/filesource/exhibits/09Q1MTPressRelease.pdf.

process, such as industrial segments that have not historically captured the attention of venture financiers. In short, we argue that policymakers have to view efforts to assist young firms within the context of the changing private-sector environment.

7.2   Venture Capital and Innovation

It is helpful to begin by briefly considering the role venture capital investors play. The financing of young and growing companies is a risky business. Uncertainty and informational gaps often characterize these organizations. These information problems make it difficult to assess these companies and permit opportunistic behavior by entrepreneurs after the financing is received.

7.2.1   Conflicts in the Venture Process

To briefly review the types of conflicts that can emerge in these settings, conflicts between managers and investors can affect the willingness of institutional investors to provide capital. If the firm raises equity from outside investors, the manager has an incentive to engage in wasteful expenditures (e.g., lavish offices) because he or she may benefit disproportionately from these but does not bear their entire cost. Similarly, if the firm raises debt, the manager may increase risk to undesirable levels. Because providers of capital recognize these problems, outside investors demand a higher rate of return than would be the case if the funds were internally generated.

Additional problems may appear in the types of more mature companies in which venture capital firms specializing in growth equity invest. For instance, entrepreneurs might invest in strategies or projects that have high personal returns but low expected monetary payoffs to shareholders.

Even if the manager wants to maximize firm value, information gaps may make raising external capital more expensive or even preclude it entirely. Equity offerings of companies may be associated with a "lemons" problem: that is, if the manager is better informed about the company's investment opportunities and acts in the interest of current shareholders, then he or she will only issue new shares when the company's stock is overvalued. Indeed, numerous studies have documented that stock prices decline upon the announcement of equity issues, largely because of the negative signal sent to the market. This "lemons" problem leads investors to be less willing to invest at attractive valuations in young or restructuring companies, or even to invest at all.

Specialized intermediaries, such as venture capital organizations, can address these problems. By intensively scrutinizing companies before providing capital and then monitoring them afterward, they can alleviate some of the information gaps and reduce capital constraints. Thus, it is important to understand the tools that venture capital investors use in this difficult environment, which enable companies ultimately to receive the financing that they cannot raise from other sources. It is the nonmonetary aspects of venture capital that are critical to its success: the screening of investments, the use of convertible securities, the syndication and staging of investments, and the provision of oversight and informal coaching.

7.2.2   The Tools of Venture Capital

Where, then, does the venture capital advantage come from? To address the information problems delineated in the preceding, venture capital investors employ a variety of mechanisms, which seem to be critical in boosting innovation.

The first of these is the screening process that venture capital investors use in selecting investment opportunities. This process is typically far more efficient than the process that other funders of high-risk projects, such as corporate research and development (R&D) laboratories and government grant makers, typically use. For instance, most large, mature corporations tend to look at their existing lines of business when choosing projects to fund. Technologies outside the firm's core market, or projects that raise internal political tensions, often get shelved. In fact, many successful venture-backed start-ups are launched by employees who leave when their companies decline to pursue what these employees see as a promising technology.

Numerous studies have documented that typical venture capital fund managers use an exhaustive process to assess the large number of business plans they receive each year. One of the pioneering studies (Wells 1974) described a typical process:

> 1) Conversations with venture capitalists that ask[ed firm] to look at company; 2) Checked personal references of controller, vice-president, and president; 3) Met with company's founders and controller; 4) Conversation with loan officer at major insurance company. The insurance company's loan committee had turned down company's request for financing even though the loan officer recommended it; 5) Conversation with company's accountant . . . ; 6) Conversation with local banker who slightly knew the company; 7) Conversation with banker who handles company's account; 8) Telephone conversation with director of company; 9) Talked to about 30 users; 10) Talked to two suppliers; 11) Talked to two competitors.

One sophisticated individual investor, who follows an approach similar to independent firms, suggests it is likely to take up to 160 hours to properly screen an opportunity (Amis and Stevenson 2001, 114). A leading venture capital group, Bessemer Venture Partners, prepared a "Due Diligence Booklet" that all potential investors were supposed to complete for each investment. This fifty-page publication raised a large variety of questions about the industry, the company, the people, and the transaction itself.

How do venture capital investors make sense of all the data they gather during this assessment process? Clearly, certain measures are more important than others. After interviewing a large number of funds about their investment criteria, Tyebjee and Bruno (1984) described the most common criteria as follows:

1. Market attractiveness (size, growth, and access to customers)
2. Product differentiation (uniqueness, patents, technical edge, profit margin)
3. Managerial capabilities (skills in marketing, management, finance, and the references of the entrepreneur)
4. Environmental threat resistance (technology life cycle, barriers to competitive entry, insensitivity to business cycles, and downside risk protection)
5. Cash-out potential (future opportunities to realize capital gains by merger, acquisition, or public offering)

Steve Kaplan and Per Strömberg (2004), who examined the actual analyses that the venture capital funds undertake when presenting potential transactions to their investment committees, identify a similar set of findings. They grouped the key decision-making criteria into three overall categories: (a) internal factors (quality of management, performance to date, funds at risk, influence of other investors, portfolio fit, and monitoring costs and valuation); (b) external factors (market size and growth, competition and barriers to entry, likelihood of customer adoption, and financial market and exit conditions); and (c) difficulty of execution (nature of the product or technology and the business strategy model).

Another way in which venture capital investors screen transactions is through financial analyses. They carefully analyze what the prospective returns from these investments will be, conditional on the firm being successful. They only invest if the expected return is suitably high. This requirement of a very high return, if the firm is successful, stems from the high failure rates associated with early-stage and restructuring investments. For instance, only approximately one-third of venture capital-backed firms complete initial public offerings, typically the most attractive route in which to exit investments. While some investments are exited successfully though acquisitions, in most cases, these investments generate far lower returns. Even in later-stage investing, the frequency with which things do not go according to plan leads to demands for high hurdle rates. Despite all the care and expertise of venture capital investors, disappointment is the rule rather than the exception.

In addition to the careful interviews and financial analysis, venture capitalists will often make investments with other investors. One firm will originate the deal and look to bring in other firms. Involving other venture capital firms provides a second opinion on the investment opportunity. There is usually no clear-cut answer as to whether any of the investments that a venture capital organization undertakes will yield attractive returns. Having other investors approve the deal limits the danger that bad deals will get funded. This is particularly true when the company is early-stage or operating in an uncertain market. Syndication also allows the venture capital firm to diversify. If the venture capital investor had to invest alone into all the companies in his portfolio, then he or she could make far fewer investments. By syndicating investments, the venture capital investor can invest in more projects and largely diversify away firm-specific risk.

The result of this detailed analysis is, of course, a lot of rejections: studies suggest only 1/2 to 1 percent of business plans seem to be funded (Wells 1974; Fenn, Liang, and Prowse 1996). Inevitably, many good ideas are rejected as part of this process. Most venture capital investors are embarrassed to admit these goofs, but Bessemer cheerily posts their "antiportfolio" of great companies they passed on for various reasons.[3] And, of course, many companies are funded, which ultimately prove to be disappointments.

When venture capital investors invest, they typically hold not common stock, but rather preferred stock. The significance of this distinction is that if the company is liquidated or otherwise returns money to the shareholders, the preferred stock will get paid before the common stock that the entrepreneurs, as well as other, less-privileged investors, hold. Moreover, venture capital investors add numerous restrictive covenants and provisions to the preferred stock. They may be able, for instance, to block future financings if the valuation is not what they are comfortable with, replace the entrepreneur, and have a set number of representatives on (or even control of) the board of directors. In this way, if something unexpected happens (which is the rule rather than the exception with entrepreneurial and restructuring firms), the venture capital investor can assert control. These terms vary with the financing round, with the most onerous terms reserved for the earliest financing rounds.

In addition to the initial selection process, the advice that venture capital firms provide to entrepreneurs, as well as the post-investment monitoring and control, support top-quality innovation. Venture capital investors also tend to spot more potential future applications of technology and business models than larger, mature companies do, perhaps because older companies focus on narrower markets.

The staging of investments also improves the efficiency of venture capital funding. In large corporations, R&D budgets are typically set out at the beginning of a project, with few interim reviews planned. Even if projects do get reviewed midstream, few of them are terminated when signs suggest they're not working out.

These practices contrast with the venture capital and growth equity pro-

---

3. See http://www.bvp.com/Portfolio/AntiPortfolio.aspx.

cess: once the decision to invest is made, these venture capital investors frequently disburse funds in stages. The refinancing of these firms, termed "rounds" of financing, is made conditional on achieving certain technical or market milestones. Providing financing in this fashion allows the venture capital investor to gather more information before providing additional funding, thus helping investors begin to separate which investments are likely to be successful and which are likely to fail. Managers of the venture- and growth equity-backed firms have to return repeatedly to their financiers for additional capital, which allows the venture capitalists to ensure that the money is not squandered on unprofitable projects. Thus, an innovative idea only continues to be funded if its promoters are able to continue to execute, and, conversely, those projects that prove promising are able to access capital in a timely fashion.

Finally, venture capital investors also provide intensive oversight of the firms. Michael Gorman and Bill Sahlman (1989) found that venture capital investors who responded to their survey spent about half their time monitoring an average of nine portfolio investments and serving on the boards for five of those nine companies. They visited their companies relatively frequently and spent an average of eighty hours a year on site with the company on whose board they served. Frequent telephone conversations amounted to another thirty hours per year for each company. In addition, they worked on the company's behalf by attracting new investors, evaluating strategy against new conditions, and interviewing/recruiting new management candidates.

Interviews with venture capital investors and entrepreneurs suggest that the consequence of these tools is that venture capital investors play an important role in boosting the firms that they fund. Their assistance has several dimensions: accelerating growth, professionalizing and improving management practices, and ensuring long-run success (see, for instance, Gurung and Lerner 2008, 2009).

What prohibits other financial intermediaries (e.g., banks) from undertaking the same sort of monitoring? While it is easy to see why individual investors may not have the expertise to address these types of agency problems, it might be thought that bank credit officers could undertake this type of oversight. Yet even in countries with exceedingly well-developed banking systems, such as Germany and Japan, policymakers today are seeking to encourage the development of a venture capital industry to insure more adequate financing for risky entrepreneurial companies. The limitations of banks stem from several of their key institutional features, from regulations, skill sets, to compensation schemes.

7.2.3 Large-Sample Evidence

Clearly, venture capital exerts a major impact on the fates of individual companies. But does all this fundraising and investing influence the overall economic landscape as well? How could it even be determined whether such an influence exists? And if it did exist, how would it be measured?

To assess this question, we can look at studies of the experience of the market with the most-developed and seasoned venture capital industry, the United States. Despite the fact that venture activity is particularly well developed in this nation, the reader might be skeptical as to whether this activity would noticeably impact innovation: for most of past three decades, investments made by the entire venture capital sector totaled less than the R&D and capital-expenditure budgets of large, individual companies such as IBM, General Motors, or Merck. On the face of it, this suggests the business press has exaggerated the importance of the venture capital industry. After all, high-tech start-ups make for interesting reporting, but do they really redefine the U.S. economy?

One way to explore this question is to examine the impact of venture investing on wealth, jobs, and other financial measures across a variety of industries. Though it would be useful to track the fate of *every* venture capital-financed company and find out where the innovation or technology ended up, in reality, only those companies that have gone public can be tracked. Consistent information on venture-backed firms that were acquired or went out of business simply doesn't exist. Moreover, investments in companies that eventually go public yield much higher returns than support given to firms that get acquired or remain privately held.

These firms have had an unmistakable effect on the U.S. economy. In September 2008, 895 firms were publicly traded on U.S. markets after receiving their private financing from venture capitalists (this does not include the firms that went public but were subsequently acquired or delisted). One way to assess the overall impact of the venture capital industry is to look at the economic "weight" of venture-backed companies in the context of the larger economy.[4] By late 2008, venture-backed firms that had gone public made up over 13 percent of the total number of public firms in existence in the United States at that time. And of the total market value of public firms ($28 trillion), venture-backed companies came in at $2.4 trillion—8.4 percent.

Venture-funded firms also made up over 4 percent (nearly one trillion dollars) of total sales ($22 trillion) of all U.S. public firms at the time. And contrary to the general perception that venture-supported companies are not profitable, operating income margins for these companies hit an average of 6.8 percent—close to the average public-company profit margin of 7.1 percent. Finally, those public firms supported by venture funding employed 6 percent of the total public-company workforce—most of these jobs high-salaried, skilled positions in the technology sector. Clearly, venture investing fuels a substantial portion of the U.S. economy.

4. This analysis is based on the author's tabulation of unpublished data from Securities Data Company (SDC) Venture Economics, with supplemental information from Compustat and the Center for Research into Securities Prices (CRSP) databases.

Venture investing not only supports a substantial fraction of the U.S. economy, but it also strengthens particular industries. To be sure, it has relatively little impact on industries dominated by mature companies—such as the manufacturing industries. That's because venture investors' mission is to capitalize on revolutionary changes in an industry, and the preceding sectors often have a relatively low propensity for radical innovation.

But contrast those industries with highly innovative ones, and the picture looks completely different. For example, companies in the computer software and hardware industry that received venture backing during their gestation as private firms represented more than 75 percent of the software industry's value. Venture-financed firms also play a central role in the biotechnology, computer services, and semiconductor industries. All of these industries have experienced tremendous innovation and upheaval in recent years. Venture capital has helped catalyze change in these industries, providing the resources for entrepreneurs to generate substantial return from their ideas. In recent years, the scope of venture groups' activity has been expanding rapidly in the critical energy and environmental field, though the impact of these investments remains to be seen.

As these statistics suggest, venture capitalists create whole new industries and seed fledgling companies that later dominate those industries. The message is clear: the venture capital revolution served as the driving force behind the transformation of the U.S. economy in recent decades.

It might be thought that it would be not difficult to address the question of the impact of venture capital on innovation. For instance, one could seek to explain across industries and time whether, controlling for R&D spending, venture capital funding has an impact on various measures of innovation. But even a simple model of the relationship between venture capital, R&D, and innovation suggests that this approach is likely to give misleading estimates.

This is because both venture funding and innovation could be positively related to a third unobserved factor, the arrival of technological opportunities. Thus, there could be more innovation at times that there was more venture capital, not because the venture capital caused the innovation, but rather because the venture capitalists reacted to some fundamental technological shock that was sure to lead to more innovation. To date, only two papers have attempted to address these challenging issues.

The first of these papers, by Thomas Hellmann and Manju Puri (2000), examines a sample of 170 recently formed firms in Silicon Valley, including both venture-backed and nonventure firms. Using questionnaire responses, they find evidence that venture capital financing is related to product market strategies and outcomes of start-ups. They find that firms that are pursuing what they term an innovator strategy (a classification based on the content analysis of survey responses) are significantly more likely and faster to obtain venture capital. The presence of a venture capitalist is also associated with a significant reduction in the time taken to bring a product to market, especially for innovators (probably because these firms can focus more on innovating and less on raising money). Furthermore, firms are more likely to list obtaining venture capital as a significant milestone in the life cycle of the company as compared to other financing events.

The results suggest significant interrelations between investor type and product market dimensions and a role of venture capital in encouraging innovative companies. But this does not definitively answer the question of whether venture capitalists cause innovation. For instance, we might observe personal injury lawyers at accident sites, handing out business cards in the hopes of drumming up clients. But just because the lawyer is at the scene of the car crash does not mean that he caused the crash. In a similar vein, the possibility remains that more innovative firms choose to finance themselves with venture capital, rather than venture capital causing firms to be more innovative.

In my work with Sam Kortum, I visit the same question. Here, we look at the aggregate level: did the participation of venture capitalists in any given industry over the past few decades lead to more or less innovation? It might be thought that such an analysis would have the same problem as the preceding personal injury lawyer story. Put another way, even if we see an increase in venture funding and a boost in innovation, how can we be sure that one caused the other?

We address these concerns about causality by looking back over the industry's history. In particular, as we discussed in the preceding, a major discontinuity in the recent history of the venture capital industry was the U.S. Department of Labor's clarification of the Employee Retirement Income Security Act (ERISA) in the late 1970s, a policy shift that freed pensions to invest in venture capital. This shift led to a sharp increase in the funds committed to venture capital. This type of external change should allow us to figure out what the impact of venture capital was because it is unlikely to be related to how many or how few entrepreneurial opportunities there were to be funded.

Even after addressing these causality concerns, the results suggest that venture funding does have a strong positive impact on innovation. The estimated coefficients vary according to the techniques employed, but, on average, a dollar of venture capital appears to be *three to four* times more potent in stimulating patenting than a dollar of traditional corporate R&D. The estimates, therefore, suggest that venture capital, even though it averaged less than 3 percent of corporate R&D in the United States from 1983 to 1992, is responsible for a much greater share—perhaps 10 percent—of U.S. industrial innovations in this decade.

A natural worry with the preceding analysis is that it looks at the relationship between venture capital and patenting, not venture capital and innovation. One possible explanation is that such funding leads entrepreneurs to protect their intellectual property with patents rather than other mechanisms such as trade secrets. For instance, it may be that the entrepreneurs can fool their venture investors by applying for large number of patents, even if the contributions of many of them are very modest. If this is true, it might be inferred that the patents of venture-backed firms would be lower quality than nonventure-backed patent filings.

How could this question of patent quality be investigated? One possibility is to check the number of patents that cite a particular patent.[5] Higher-quality patents, it has been shown, are cited by other innovators more often than lower-quality ones. Similarly, if venture-backed patents are lower quality, then companies receiving venture funding would be less likely to initiate patent-infringement litigation. (It makes no sense to pay money to engage in the costly process of patent litigation to defend low-quality patents.)

So what happens when patent quality is measured with these criteria? As it happens, the patents of venture-backed firms are more frequently cited by other patents and are more aggressively litigated—thus, it can be concluded that they are high quality. Furthermore, the venture-backed firms more frequently litigate trade secrets, suggesting that they are not simply patenting frantically in lieu of relying on trade-secret protection. These findings reinforce the notion that venture-supported firms are simply more innovative than their nonventure-supported counterparts.

Mollica and Zingales (2007), by way of contrast, focus on regional patterns: as a regional unit, they use the 179 Bureau of Economic Analysis economic areas, which are composed by counties surrounding metropolitan areas. They exploit the regional, cross-industry, and time series variability of venture investments in the United States to study the impact of venture capital activity on innovation and the creation of new businesses. Again, they grapple with causality issues by using an instrumental variable: as an instrument for the size of venture capital investments, they use the size of a state pension fund's assets. The idea is that state pension funds are subject to political pressure to invest some of their funds in new businesses in the states. Hence, the size of the state pension fund triggers a shift in the local supply of venture capital investment, which should help identify the effect of venture capital on patents.

Even with these controls, they find that venture capital investments have a significant positive effect both on the production of patents and on the creation of new businesses. A one standard deviation increase in the venture capital investment per capita generates an increase in the number of patents between 4 and 15 percent. An increase of 10 percent in the volume of venture capital investment increases the total number of new businesses by 2.5 percent.

---

5. Patent applicants and examiners at the patent office include references to other relevant patents. These serve a legal role similar to that of property markers at the edge of a land holding.

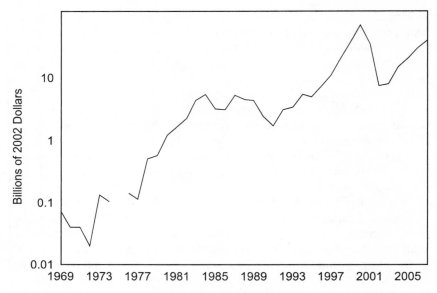

Fig. 7.1   Venture capital fundraising by year, 1969–2007

Source: The figure is based on unpublished Asset Alternatives and Venture Economics databases.

Note: There was no venture fundraising in 1975.

7.3   Cyclicality in the Venture Capital Industry

But venture capital also is far from a seamless and steady way to fund innovation, as our opening discussion suggested. The recent changes in the venture capital market are not the first such cycles in the venture market. Figures 7.1 and 7.2 depict the changing amount of venture capital funds raised and the returns from these funds. In this section, we will explore what accounts for such extreme variations.

7.3.1   A Simple Framework

To help understand the dynamics of the venture capital industry, it is helpful to employ a simple framework.[6] The two critical elements for understanding shifts in venture capital fundraising are straightforward: a demand curve and a supply curve. Just as in markets for commodities like oil and semiconductors, shifts in supply and demand shape the amount of capital raised by venture funds. These also drive the returns that investors earn in these markets.

The supply of venture capital is determined by the willingness of investors

6. The supply and demand framework for analyzing venture capital discussed here was introduced in Poterba (1989) and refined in Gompers and Lerner (1998).

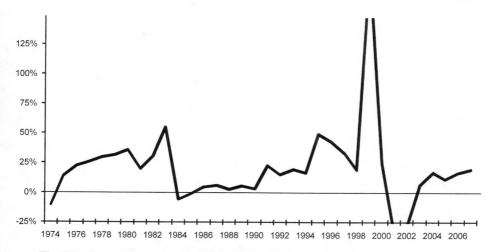

Fig. 7.2   Returns to venture capital investments, 1974–2007
Source: The figure is based on an unpublished Venture Economics database.

to provide funds to venture firms. The willingness of investors to commit money to venture capital funds, in turn, is dependent upon the expected rate of return from these investments relative to the return they expect to receive from other investments. Higher expected returns lead to a greater desire of investors to supply venture capital. As the return that investors expect to earn from their venture investments increases—that is, as we go up the vertical axis—the amount supplied by investors grows (we move further to the right column, the horizontal axis).

The number of entrepreneurial firms seeking venture capital determines the demand for capital. Demand is also likely to vary with the rate of return anticipated by investors. As the minimum rate of return sought by the investors increases, fewer entrepreneurial firms can meet that threshold. The demand schedule typically slopes downward: higher return expectations lead to fewer financeable firms because fewer entrepreneurial projects can meet the higher hurdle.

Together, supply and demand should determine the level of venture capital in the economy. This is illustrated in figure 7.3. The level of venture capital should be determined by where the two lines—the supply curve (*S*) and the demand curve (*D*)—meet. Put another way, we would expect a quantity *Q* of venture capital to be raised in the economy, while the funds to earn a return of *R*, on average.

It is natural to think of supply and demand curves as smooth lines. But this is not always the case. Consider, for instance, the venture capital market before Department of Labor's clarification of the "prudent man" rule of the ERISA in 1979. The willingness of investors to provide capital before the

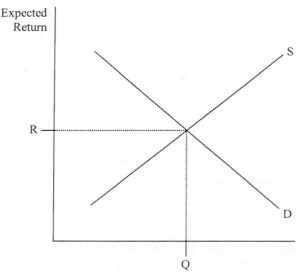

Fig. 7.3   Steady-state level of venture capital

clarification of ERISA policies looked like the supply curve may be been distinctly limited: no matter how high the expected rate of return for venture capital was, the supply would be limited to a set amount. The vertical segment of the supply curve resulted because pension funds, a segment of the U.S. financial market that controlled a substantial fraction of the long-term savings, were simply unable to invest in venture funds. Consequently, the supply of venture capital may have been limited at any expected rate of return.

7.3.2   The Impact of Shifts

These supply and demand curves are not fixed. For instance, the shift in ERISA policies led to the supply of funds moving outward. Similarly, major technological discoveries, such as the development of genetic engineering, led to an increase in the demand for venture capital.

But the quantity of venture capital raised and the returns it enjoys often do not adjust quickly and smoothly to the changes in supply and demand curves. We can illustrate this by comparing the venture capital market to that for snack foods. Companies like Frito-Lay and Nabisco closely monitor the shifting demand for their products, getting daily updates on the data collected in supermarket scanners. They restock the shelves every few days, adjusting the product offering in response to changing consumer tastes. They can address any imbalances of supply and demand by offering coupons to consumers or making other special offers.

By way of contrast, in the venture market, the quantity of funds provided may not shift rapidly. The adjustment process is often quite slow and uneven, which can lead to substantial and persistent imbalances. When the quantity provided does react, the shift may "overshoot" the ideal amount and lead to yet further problems.

This can be illustrated again using our framework. It is important to distinguish here between short-and long-run curves. While in the long run, the curve may have a smooth upward slope, the short-run curve may be quite different. The long-run supply curve (*SL*) may have a smooth upward slope. But the supply in the short run may be essentially fixed if investors cannot or will not adjust their allocations to venture capital funds. Thus, the short-run curve may instead be a vertical line (*SS*).

This difference is illustrated figure 7.4, which explores the short- and long-run impact of a positive demand shock. The discovery of a new scientific approach, such as genetic engineering, or the diffusion of a new technology, such as the transistor or the Internet, may have a profound effect on the venture capital industry. As large companies struggle to adjust to these new technologies, numerous agile small companies may seek to exploit the opportunity. As a result, for any given level of return demanded by investors, there now may be many more attractive investment candidates.

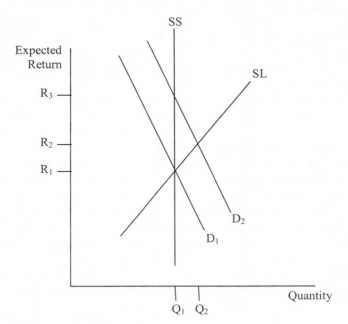

Fig. 7.4   Impact on quantity of a demand shock

In the long run, the quantity of venture capital provided will adjust upward from Q_1 to Q_2. Returns will also increase, from R_1 to R_2. In the months or even years after the shock, however, the amount of venture capital available may be essentially fixed. Instead of leading to more companies being funded, the return to the investors may climb dramatically, up to R_3. Only with time will the rate of return gradually subside as the supply of venture capital adjusts.

There are at least two factors that might lead to such short-run rigidities. These are the structure of the funds themselves and the slowness with which information on performance is reported back to investors. We will explore how each factor serves to dampen the speed with which the supply of venture capital adjusts to shifts in demand.

The Nature of Venture Funds

When investors wish to increase their allocation to public equities or bonds, this change is easily accomplished. These markets are "liquid": shares can be bought and sold easily, and adjustments in the level of holdings can be readily accomplished. The nature of venture capital funds, however, makes these kind of rapid adjustments much more difficult.

Consider an instance where a university endowment decides that venture capital is a particularly attractive investment class and decides to increase its allocation to these investments. From the time at which this new target is agreed upon, it is likely to be several years before the policy is fully implemented. Because venture funds only raise funds every two or three years, if the endowment simply wants to increase its commitment to existing funds, they will need to wait until the next fundraising cycle occurs for these funds. In many cases, they may be unable to invest as much in the new funds as they wish.

The reluctance of venture groups to accept their capital stems from the fact that the number of experienced venture capitalists often adjusts more slowly than the swings in capital. Many of the crucial skills of being an effective venture capitalist cannot be taught formally: rather, they need to be developed through a process of apprenticeship. Furthermore, the organizational challenges associated with rapidly increasing the size of a venture partnership are often wrenching ones. Thus, groups such as Kleiner Perkins and Greylock have resisted rapidly increasing their size even if investor demand is so great that they could easily raise many billions of dollars.

If, indeed, the endowment decides to undertake a strategy of investing in new funds, potential candidates for the university's funds will need to be exhaustively reviewed. Once the funds are chosen, the investments will not be made immediately. Rather, the capital that the university commits will only be drawn down in stages over a number of years.

The same logic works in reverse. If the endowment or pension officers decide to scale back their commitment to private equity, it is likely to take a number of years to do so. An illustration of this stickiness was seen following the stock market correction of 1987. Many investors, noting the extent of equity market volatility and the poor performance of small high-technology stocks, sought to scale back their commitments to venture capital. Despite the correction, flows into venture capital funds continued to rise, not reaching their peak until the last quarter of 1989.[7]

Another contributing factor is self-liquidating nature of venture funds. When venture funds exit investments, they do not reinvest the funds but, rather, return the capital to their investors. These distributions are typically either in the form of stock in firms that have recently gone public or cash. The pace of distributions varies with the rate at which venture capitalists are liquidating their holdings.

Thus, during "hot" periods with large numbers of initial public offerings and acquisitions—which are likely to be the times when many investors desire to increase their exposure to venture capital—limited partners receive large outflows from venture funds. Even to maintain the same percentage allocation to venture funds during these peak periods, the institutions and individuals must accelerate their rate of investment. Increasing their exposure is consequently quite difficult. Conversely, during "cold" periods, when investors are likely to wish to reduce their allocation to this asset class, they receive few distributions. Thus, it is often difficult to achieve a desired exposure to venture capital during periods of rapid change in the market.

The Role of Information Lags

A second factor contributing to the stickiness of the supply of venture capital is the difficulty in discerning what the current status of the venture market is at any given time. While mutual and hedge funds holding public securities are "marked to market" on a daily basis, the delays between the inception of a venture investment and the discovery of its quality is long indeed.

The information lags can have profound effects. For instance, when the investment environment becomes far more attractive, it can take a number of years to fully realize the fact. While investments in Internet-related securities in the mid-1990s yielded extremely high returns, it took many years for the bulk of institutional investors to realize the size of the opportunity. Similarly, when the investment environment becomes substantially less attractive, as it did during the spring of 2000, investors often continue to plough money into funds. (see, for instance, the discussion in Kreutzer [2001].)

Some of these information problems stem from the firms themselves. The types of firms that attract venture capital are surrounded by substantial uncertainty and information gaps. But these inevitable difficulties are exacerbated by the manner in which the performance of funds is typically reported.

7. This claim is based on an analysis of an unpublished Venture Economics database.

The first of these is the conservatism of the valuations. Venture groups tend to be extremely conservative in reporting how much the firms they invest in are worth, at least until the firms are taken public or acquired. While this limits the danger that investors will be misled into thinking that the funds is doing better than it actually is, this practice minimizes the information flow about the current state of the market.[8]

This reporting practice, for instance, must lead us to be cautious in evaluating the returns depicted in figure 7.2. Because relatively few firms get taken public during "cold" markets and many do during "hot" ones, there are many more dramatic write-ups in firms during the years with active public markets. But the actual value-creation process in venture investments is quite different. In many cases, the value of a firm actually increases gradually over time, even as it is being held at cost. Thus, the low returns during cold periods understate the progress that is being made, just as the high returns during the peak periods overstate the success during those years. Thus, the signals that venture groups receive are quite limited.

An Illustration

The preceding discussion ignores many of the complex institutional realities that affect the ebbs and flows of venture capital fundraising. But even such simple tools can be quite helpful in understanding overall movements in the venture capital activity, as can be illustrated by considering the recent history of the venture capital industry.

As figure 7.1 illustrates, the supply of venture funding began growing rapidly in the mid-1990s. Many practitioners at the time viewed this event glumly, arguing that a boost in venture activity must inevitably lead to a deterioration of returns. Yet the investments during this period enjoyed extraordinary success, as figure 7.2 illustrates. How could these seasoned observers have been so wrong?

The reason is that these years saw a dramatic shift in the opportunities available to venture capital investors. The rapid diffusion of Internet access and the associated development of the World Wide Web ushered in an extraordinary period in the U.S. economy. The ability to transfer visual and text information in a rapid and interactive manner was a powerful tool, one that would transform both retail activities as well as the internal management of firms.

Such a change led to an increase in the demand for venture capital financing. Thus, for any given level of return that investors demanded, there should have been a considerably greater number of opportunities to fund. Far from declining, the rate of return that venture investments enjoyed actually rose. Much of this rise reflected the fact that the supply of effective and cred-

---

8. The problems with the accounting schemes used by venture capital groups are discussed in Cain (1997), Gompers and Lerner (1997), and Reyes (1990).

ible venture organizations adjusted only slowly. As a result, those groups who were active in the market during this period enjoyed extraordinary successes.

7.3.3   Why Does the Venture Market Overreact?

Another frequently discussed pathology in the venture market is the other side of the same coin. Once the markets do adjust to the changing demand conditions, they frequently go too far. The supply of venture capital ultimately will rise to meet the increased opportunities, but these shifts often are too large. Too much capital may be raised for the outstanding amount of opportunities. Instead of shifting to the new steady-state level, the short-term supply curve may shift to an excessively high level.

The same problem can occur in reverse. A downward shift in demand can trigger a wholesale withdrawal from venture capital financing. Returns rise dramatically as a result. While the supply of venture capital will ultimately adjust, in the interim, promising companies may not be able to attract funding. In this section, we explore two possible explanations for this phenomenon.

Do Public Markets Provide Misleading Information?

One possibility is that institutional investors and venture capitalists may overestimate the shifts that have occurred. They may believe that there are tremendous new opportunities and, consequentially, shift the supply of venture capital to meet that apparent demand.

This suggestion is captured in figure 7.5. A positive shock to the demand for venture capital occurs, moving the demand curve out from D_1 to D_2. Limited and general partners, however, mistakenly believe that the curve has shifted out to D_3. The short-run supply curve thus shifts from SS_1 to SS_3, leaving excessive investment and disappointing returns in its wake.

Such mistakes may arise because of misleading information from the public markets. Examples abound where venture capitalists have made substantial investments in new sectors, at least partially responding to the impetus provided by the high valuations in that sector. Understanding why public markets overvalue particular sectors is beyond the scope of this piece. Certainly, though, it seems in some cases that investors fail to take into account the impact of competitors: firms appear to be valued as if they are the sole firm active in a sector, and the impact of competitors on revenues and profit margins are not fully anticipated.

Whatever the causes of these misvalautions, historical illustrations are plentiful. One famous example was during the early 1980s, when nineteen disk drive companies received venture capital financing. (For detailed discussions, see Sahlman and Stevenson [1985] and Lerner [1997].) Two-thirds of these investments came in 1982 and 1983, as the valuation of publicly traded computer hardware firms soared. Many disk drive companies also

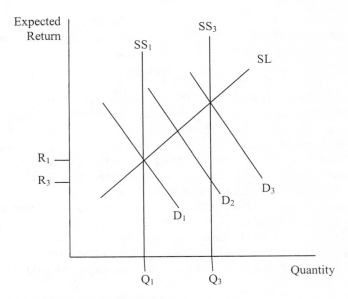

Fig. 7.5   Misleading public market signals

went public during this period. While industry growth was rapid during this period of time (sales increased from $27 million in 1978 to $1.3 billion in 1983), it was questioned at the time whether the scale of investment was rational given any reasonable expectations of industry growth and future economic trends. Indeed, between October 1983 and December 1984, the average public disk drive firm lost 68 percent of its value. Numerous disk drive manufacturers that had yet to go public were terminated, and venture capitalists became very reluctant to fund computer hardware firms.

Unreasonable swings in the public markets may also lead to over- and underinvestment in venture capital as a whole. Institutions typically try to keep a fixed percentage of their portfolio invested in each asset class. Thus, when public equity values climb, institutions are likely to want to allocate more to venture capital. If the high valuations are subsequently revealed to be without foundations, the level of venture capital will have once again overshot its target.

Do Venture Capitalists Underestimate the Cost of Change?

A second explanation for the "overshooting" phenomenon is venture capitalists' failure to consider the costly adjustments associated with the growth of their own investment activity. The very act of growing the pool of venture capital under management may cause distractions and introduce organizational tensions. Even if demand has expanded, the number of opportunities that a venture group—or the industry as a whole—can address may at first be limited.

Why might these adjustment costs come about? One possibility is that growth frequently leads to changes in the way in which venture groups invest their capital, which has a deleterious effect on returns. A second possibility is that growth introduces strains on the venture organization itself.

First, consider the types of pressures that rapid growth imposes on the venture investment process. Rather than making more investments, rapidly growing venture organizations frequently attempt to increase their average investment size. In this way, the same number of partners can manage a larger amount of capital without an increase in the number of firms that each needs to scrutinize. This shift to larger investments has frequently entailed making larger capital commitments to firms up-front. This has the potential cost of reducing the venture capitalist's ability to control the firm using staged capital commitments.

Similarly, venture firms syndicate less with their peers during these times. By not syndicating, venture groups can put more money to work. As the sole investor, the venture groups can allow each of its partners to manage more capital while keeping the number of companies that he or she is responsible for down to a manageable level. But this syndication can have a number of advantages, such as helping reduce the danger of costly investment mistakes.

Another set of explanation factors relates to organizational pressures. Limited and general partners may underestimate the consequences of expanding the scale (and the scope) of the fund. An essential characteristic of venture capital organizations has been the speed with which decisions can be made and the parallel incentives that motivate the parties. An expansion of the fund can lead to a fragmentation of the bonds that tie the partnership into a cohesive whole.

One dramatic illustration of these challenges is the experience of Schroder Ventures (Bingham, Ferguson, and Lerner 1996). Schroders' private equity effort began in 1985 with funds focused on British venture capital and buyout investments. Over time, however, they added funds focusing on other markets, such as France and Germany, and particular technologies, such as the life sciences. The venture capitalists—and the institutional investors backing them—realized that there were substantial opportunities in these other markets.

But as the venture organization grew, substantial management challenges emerged. In particular, it became increasingly difficult to monitor the investment activities of each of the groups, a real concern because the parent organization served as the general partner of each of the funds (and, thus, was ultimately liable for any losses). Each of the groups saw itself as an autonomous entity, and even in some cases resisted cooperating (and sharing the capital gains) with the others. While the organization eventually completed a restructuring that allowed it to raise a single fund for all of Europe, the process of change was a slow and painful one.

These tensions are by no means confined to international venture capital organizations. Very similar tensions have appeared in U.S. rapidly growing groups between general partners specializing in life science and information technology and those located in different regions. In some instances, one of these groups has become convinced that the other is getting a disproportionate share of rewards in light of their relative investment performances. In others, it has become difficult to coordinate and oversee activities.

In some cases, these tensions have led to groups splitting apart. For instance, in August 1999, Institutional Venture Partners and Brentwood Venture Capital—venture funds that had each invested about one billion dollars over several decades—announced their intention to restructure (Barry and Toll 1999). The information technology and life sciences venture capitalists from the two firms indicated that they would join with each other to form two new venture capital firms. Pallidium Venture Capital would exclusively pursue health care transactions, while Redpoint Ventures would focus on Internet and broadband infrastructure investments. Press accounts suggested the decision was largely driven by the dissatisfaction of some of the information technology partners at the firms, who felt that their stellar performance had not been appropriately recognized.

In other cases, a key partner—often dissatisfied with his or her role or compensation—has departed a venture group, entailing a real disruption to the organization. For instance, Ernest Jacquet left to form Parthenon Ventures shortly after Summit Partners closed on a $1 billion buy-out fund ("Summit's Jacquet . . ." 1998). While it is very rare for investors to demand that their funds be returned—though, for instance, Foster Capital Management returned $200 million after the several junior partners departed in 1998—these defections can, nonetheless, affect the workings and continuity of these groups ("Foster Management . . ." 1998).

In short, rapid growth puts severe pressures on venture capital organizations. Even when the problems do not result in an extreme outcome such as a group dissolving, the demands on the partners' time in resolving these problems have often been substantial. Thus, during periods of rapid growth, venture capital groups may correctly observe that there are many more opportunities to fund. Rapidly expanding to address these opportunities may be counterproductive, however, and lead to disappointing returns.

7.4   Venture Capital and Alternative Energy

Before turning to the implications for policymakers, it is worth noting that in past few years, there has been a classic boom in venture investment in alternative energy. This has reflected the fact that the cost of energy has been very high. Despite earlier disappointments, in recent years, investment has surged in the sector, and the environment has become one that is more favorable for ventures in clean energy development.

From 2004 to 2007, investment in clean energy saw a huge surge, increas-

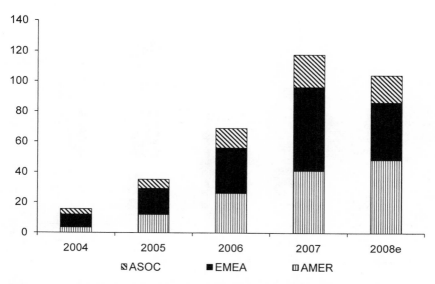

Fig. 7.6   Clean energy investment ($ in billions) by geography, 2004–2008
Source: World Economic Forum (2009).

Notes: Totals are extrapolated values based on disclosed deals from the New Energy Finance Industry Intelligence database. They do not include R&D or small projects, which is why the total in this chart is lower than the total new investment shown in other charts. ASOC-Asia Oceania region; EMEA-Europe Middle East Africa region; AMER-Americas region.

ing nearly fivefold. As recently as five years ago, venture investment in clean energy meant wind projects, mostly in Denmark, Germany, and Spain (see figure 7.6). The four-year surge in investment activity spanned all sectors, all geographies, and all asset classes, and as a result, the clean energy financing spectrum is well-developed, from very early stage investment in emerging technologies, right through to large established companies raising money on the public markets (figures 7.7 and 7.8). In the United States, investment in clean energy projects has grown dramatically in the past decade, surpassing the $13 billion mark in 2007.

This has been true not just for large firms, but also for new ventures. In 2007 alone, venture capital investment in clean energy technology companies was $2.5 billion, up from $30 million in 2001.[9]

It is important to note, however, that while the 2008 total is down only slightly from 2007 ($142 billion compared to the $148 billion in the previous year), a strong start may disguise a much weaker second half of the year due to the impact of the global financial crisis. In 2008, approximately 80 percent ($104 billion) of funding was provided by third-party investors such as venture capital and private equity, asset managers, and banks.

9. See U.S. Department of Energy (2008).

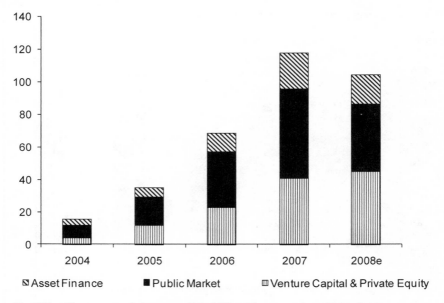

Fig. 7.7 Clean energy investment ($ in billions) by asset class, 2004–2008

Source: World Economic Forum (2009).

Notes: Totals are extrapolated values based on disclosed deals from the New Energy Finance Industry Intelligence database. They exclude R&D and small projects.

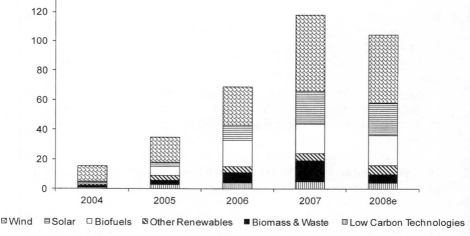

Fig. 7.8 Clean energy investment ($ in billions) by sector, 2004–2008

Source: World Economic Forum (2009).

Notes: Totals are extrapolated values based on disclosed deals from the New Energy Finance industry Intelligence database. They exclude R&D and small projects. Other Renewables includes geothermal and mini-hydro; Low Carbon Technologies includes energy efficiency fuels and power storage.

In 2005, wind was the dominant sector attracting venture capital investment. In 2006, biofuels attracted the highest venture capital investment, with the solar sector attracting the second highest amount. In 2007, 21 gigawatts of new wind capacity were added worldwide, an amount to half of new renewable energy capacity and over 11 percent of all new power generation capacity. Solar energy is now the fastest-growing sector and is a leader for venture capital investment. The development of large-scale solar projects in 2007 attracted $17.7 billion project financing, nearly a quarter of all new investment (up 250 percent from previous year). (See figure 7.9.)

The surge in investments has been reflected in the double-digit returns of these projects. In venture capital investments specifically, investors in clean technologies in Europe and the United States achieved excellent returns on their investments up to mid-2008, according to the third annual European Clean Energy Venture Returns Analysis (ECEVRA), completed by New Energy Finance in collaboration with the European Energy Venture Fair. The study is based on confidential returns by investors at the end of the first half of 2008 and covers 302 clean technology portfolio companies, representing €1.77 billion of venture capital invested in clean technology since 1997. Of these, 26 have so far resulted in public listings, and 32 have been exited or partially exited via trade sale. The success rate to date has been reasonably high with a pooled gross internal rate of return (IRR) (at the portfolio company level, not the fund level) for exited deals of over 60 percent, based on the limited number of exits and with only 23 companies being

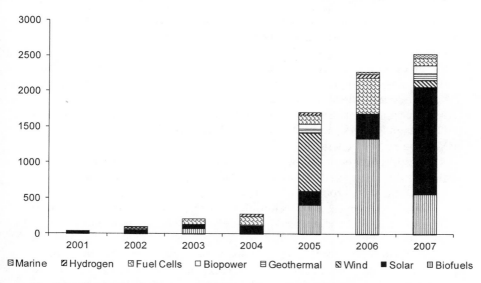

Fig. 7.9   Venture capital investment ($ in millions) in renewable energy technology companies, 2001–2007

Source: U.S. Department of Energy (2008).

Table 7.1   Comparative IRR by year of study (%)

	2006	2007	2008
Cumulative global overall IRR	49.3	43.9	61.1
Buy outs	n.d.	37.6	36.7
All venture	87.6	54.9	67.7
Europe	87.6	54.9	78.6
Europe excluding "Gorillas"	12.3	20.6	17.5
North America			7.5
All venture excluding "Gorillas"	12.3	20.6	14.5
All venture excluding "Gorillas" and recent investments held at book value	17.0	24.3	16.4

Source: New Energy Finance (2008).
Notes: n.d. = not disclosed. IRR = internal rate of return.

liquidated or written off at the time of the study. It is important to note that these exceptional returns were driven by the outstanding success of a small number of early investments in the solar sector—Q-Cells and REC in particular. Without these two particular investments, the pooled IRR was closer to 14 percent (see table 7.1). These patterns suggest some of the pattern of overfunding have been seen in this sector as well.

Looking ahead, one must consider the conditions that spawned such a surge in investment to understand the path that investment in the clean energy sector will take in the future. These extraordinary returns coincide with a new interest in all things green as well as historically cheap access to debt. The next few years will certainly be much harder for venture and private equity investors in the clean energy sector. The clean energy sector, like all other areas, has been affected by the financial crisis. Despite maintaining an estimated $14 billion of new investment thus far, excluding buy outs in 2008, and exceeding public market indices (S&P and NASDAQ), venture capital performance dropped sharply in the second quarter of 2008 (Thomson Reuters and the National Venture Capital Association [NVCA]). Venture exits in general have also fallen sharply.

Investment in the clean energy sector has suffered from three main causes. First, the industry suffered as a result of a 70 percent fall in energy prices. Second, the sector took a hit through the equity markets, as investors sold off stocks with any sort of technology or execution risk and went back to longer established businesses. Third, due to constrained credit, clean energy companies that require high capital were penalized.[10] The public markets which clean energy companies often used as a major source of fund raising, such as through initial public offerings (IPOs); secondary offerings, and convertible issues also dropped by 60 percent in 2008. (See table 7.2.)

10. See World Economic Forum (2009).

Table 7.2   **Global clean energy investment, 2007–2008, US$ billion**

Asset class	2007	2008 estimates	Change (%)
Venture capital/private equity	9.8	14.2	45
Public markets	23.4	9.4	60
Asset finance	84.5	80.6	5
Total	117.7	104.2	11

Source: World Economics Forum (2009).

Note: 2008 estimates are New Energy Finance preview figures published in October 2008.

The future of venture capital investment in the clean energy sector hinges upon two main conditions in the investing environment. The future of market conditions and the direction of government policy are two key factors in the future of clean energy sector.

Because of the nature of the higher up-front costs but lower fuel costs of clean energy projects, these projects are usually more sensitive to periods of higher interest rates or credit risk aversion. The present interest rates are a huge potential advantage for the clean energy sector. If credit markets ease—so far, banks are wary of lending capital in fear of default, and the Federal Reserve has not yet seen the results of its cheap debt—at some point, the clean energy projects could benefit tremendously, as cheap money flows into the system.

In addition, the McKinsey Global Institute notes that market and policy barriers such as lack of consumer education, fuel subsidies encouraging inefficient energy use, and asymmetrical benefits to tenants and landlords of investments in energy efficiency pose a threat.[11] As seen through the experience of Denmark and Japan, fully realizing energy efficiency opportunities requires a sustained supportive public policy. There is an opportunity to improve supply- and demand-side infrastructure. It can produce returns above cost of capital in major business. According to the report, $170 billion in energy efficient investment opportunities may have IRR of 17 percent or more.

The financial crisis also spawned changes in public policy toward the clean energy sector. For most policymakers, however, supporting the clean energy investment is seen as a way to combat the recession. In their policies of addressing urgent problems as well as long-term structural weaknesses in the economies, the clean energy sector will benefit.

If governments can lead by example, creating markets for clean energy through public procurement and mandating clean energy as well as enforcing energy efficiency standards, the investment in the clean energy sector will be bolstered and pose to be greatly profitable in the future. The suc-

11. See World Economic Forum (2009).

cess of venture investment in the clean energy sector depends heavily upon the kind of environment that governments develop for such investing. An entire ecosystem of supporting technology and service providers will be fundamental to the growth of a healthy clean energy sector—and this is inextricably linked to the ability of entrepreneurs and companies to create new businesses.[12]

7.5 The Consequences for Public Policy

While understanding the causes of cyclicality in the venture industry may be interesting, policymakers are much more likely to be interested in its consequences. In particular, to what extent do these changes affect the innovativeness of the U.S. economy?

In this section, we explore this question. We begin by considering the evidence regarding the cycles in the venture capital market on innovation. We highlight that while the overall relationship between venture capital and innovation is positive, the relationships across the cycles of venture activity may be quite different. We then consider the appropriate public policy response.

7.5.1   Innovation and Market Cycles

The evidence that venture capital has a powerful impact on innovation might lead us to be especially worried about market downturns. A dramatic fall in venture capital financing, it is natural to conclude, would lead to a sharp decline in innovation.

But this reasoning, while initially plausible, is somewhat misleading. For the impact of venture capital on innovation does not appear to be uniform. Rather, during periods when the intensity of investment is greatest, the impact of venture financing appears to decline. The uneven impact of venture on innovation can be illustrated with both case study and empirical evidence.

Field-Based Evidence

We have already discussed how in many instances the levels of funding during peak periods appear to "overshoot" the desired levels. Whether caused by the presence of misleading public market signals or the overoptimism on the part of the venture capitalists, funds appear to be deployed much less effectively during the boom period.

In particular, all too often these periods find venture capitalists funding firms that are too similar to one another.[13] The consequences of these exces-

---

12. See World Economic Forum (2009).
13. These results are also consistent with theoretical works in "herding" by investment managers. These models suggest that when, for instance, investment managers are assessed on the

sive duplication is frequently the same: highly duplicative research agendas, intense bidding wars for scientific and technical talent culminating with frequent defections from firm to firm, costly litigation alleging intellectual property and misappropriation of ideas across firms, and the sudden termination of funding for many of these concerns.

One example was the peak period of biotechnology investing in the early 1990s. While the potential of biotechnology to address human disease was doubtless substantial, the extent and nature of financing seemed to many observers at the time hard to justify. In some cases, dozens of firms pursuing similar approaches to the same disease target were funded. Moreover, the valuations of these firms often were exorbitant: for instance, between May and December 1992, the average valuation of the privately held biotechnology firms financed by venture capitalists was $70 million. These doubts were validated when biotechnology valuations fell precipitously in early 1993: by December 1993, only 42 of 262 *publicly traded* biotechnology firms had a valuation over $70 million.[14]

Most of the biotechnology firms financed during this period ultimately yielded very disappointing returns for their venture financiers and modest gains for society as a whole. In many cases, the firms were liquidated after further financing could not be arranged. In others, the firms shifted their efforts into other, less-competitive areas, largely abandoning the initial research efforts. In yet others, the companies remained mired with their peers for years in costly patent litigation.

The boom of 1998 to 2000 provides many additional illustrations. Funding during these years was concentrated in two areas: Internet and telecommunication investments, which, for instance, accounted for 39 percent and 17 percent of all venture disbursements in 1999. Once again, considerable sums were devoted to supporting highly similar firms—for example, the nine dueling Internet pet food suppliers—or else efforts that seemed fundamentally uneconomical and doomed to failure, such as companies that undertook the extremely capital-intensive process of building a second cable network in residential communities. Meanwhile, many apparently promising areas—for example, advanced materials, energy technologies, and micro manufacturing—languished unfunded as venture capitalists raced to focus on the most visible and popular investment areas. It is difficult to believe that the impact of a dollar of venture financing was as powerful in spurring innovation during these periods as in others.

---

basis of their performance relative to their peers (rather than against some absolute benchmark), they may end up making investments to similar to each other. For a review of these works, see Devenow and Welch (1996).

14. These figures are based on an analysis of an unpublished Venture Economics database.

Statistical Evidence

These suggestive accounts are borne out in a statistical analysis. Using the framework of Kortum and Lerner (2000), we show that the impact of venture capital on innovation was less pronounced during boom periods.

In this analysis, we analyze annual data for twenty manufacturing industries between 1965 and 1992. The dependent variable is U.S. patents issued to U.S. inventors by industry and date of application. Our main explanatory variables are measures of venture funding collected by Venture Economics and industrial R&D expenditures collected by the U.S. National Science Foundation (NSF).

To be sure, these measures are limited in their effectiveness. For instance, companies do not patent all commercially significant discoveries (though in the original paper, we show that the patterns appear to hold when we use other measures of innovation). Similarly, we are required to aggregate venture funding and patents into a twenty-industry scheme that is used by the NSF to measure R&D spending. Finally, our analysis must exclude the greatest boom period of all, the 1998 to 2000 surge (patent applications can only be observed with a considerable lag).

Table 7.3 presents our estimate of b, the influence of venture capital funding on patent applications, controlling for R&D spending, industry effects, and the year of the observation. Any number greater than 1 implies that venture capital is more powerful than traditional corporate R&D in spurring innovation. (This is a specification similar to regression [3.2] in that paper, with the addition of an added measure for the "hottest" periods.) We then show the implied coefficient when we estimate the impact of venture capital on innovation separately for those periods that had the great venture capital investments (defined here as the top 1 percent of industry-year observations). As the table reports, the impact of venture capital on innovation is

Table 7.3   **Implied impact of venture capital on innovation**

	Coefficient or p-value
Implied potency of venture financing, normal industry periods	13.57
Implied potency of venture financing, overheated industry periods	11.53
p-value, test of difference between normal and overheated industry periods	0.000

Source: Based on the linear patent production function estimated by Kortum and Lerner (2000).

Notes: The first row presents implied impact of venture financing on innovation for all manufacturing industries and years between 1965 and 1992 except where the levels of venture inflows are in the top 1 percent. The second row presents the implied coefficient during the industries and years where inflows are in the top 1 percent. The final row presents the *p*-value from a test that the two coefficients are identical.

some 15 percent lower during the boom periods, a difference that is strongly statistically significant.

As discussed in Kortum and Lerner (2000), the magnitude of the impact of venture capital on innovation diminishes—but remains positive and significant—when we control for reverse causality: the fact that technological breakthrough are likely to stimulate venture capital investments. When we repeat the analysis reported here using a number of these complex specifications, the magnitude of the difference between normal and boom periods remains similar, and the percentage difference widens. This statistical result corroborates the field study evidence suggesting that venture capital's impact on innovation is less pronounced during booms.

A Cautionary Note

These patterns may lead us to worry less about the short-run fluctuations in venture financing. While the impact on entrepreneurial activity is likely to be dramatic, the effects on innovation should be more modest.

This conclusion, however, must be tempered by the awareness of history: in some cases, surges in venture capital activity have been followed by pronounced and persistent downturns. As alluded to in the preceding, just as we can see "overshooting" by investors, so can we see prolonged "undershooting."

One sobering example was the 1970s. The late 1960s had seen record fundraising, both by independent venture groups and Small Business Investment Companies (SBICs), federally subsidized pools of risk capital. Many of the investments by the less-established venture groups failed in the subsequent recession, particularly those of the SBICs. (The selection process for these licenses appeared to emphasize political connections over investment acumen.) The poor returns generated a powerful reaction, leading both public and private market investors to be unwilling to contribute new capital.

Figure 7.10 depicts one consequence of the period of this reaction. The graph depicts the volume of initial and follow-on offerings in the sector that saw the greatest concentration of venture investments during this period: computer and computer-related firms. The amount of capital raised by these firms fell from $1.2 billion (in today's dollars) in 1968 to 1969 to just $201 million in the entire period from 1973 to mid-1978, with absolutely no financing being raised in many quarters. To be sure, many of the firms that raised capital during the boom years and then could not get refinanced had business plans that were poorly conceptualized or were in engaged in doomed battles with entrenched incumbents such as IBM. But many other firms seeking to commercialize many of the personal computing and networking technologies that would prove to have such a revolutionary impact in the 1980s and 1990s also struggled to raise the financing necessary to commercialize their ideas.

At the same time, it is important to note that while venture capital fund-

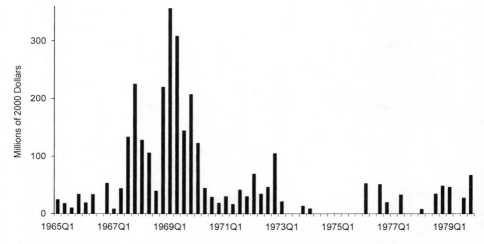

Fig. 7.10   Initial public offerings and seasoned equity offerings by computer and computer-related firms, by quarter, 1965–1979.

Sources: The authors compiled the information from *Investment Dealers' Digest,* the Securities Data Company database, and other sources.

raising and investment has cooled down considerably from the "white hot" days of 2000, the level of activity is still extremely high from a historical perspective. In fact, if we were to remove the 1999 to 2000 "bubble" period from figure 7.1, the venture industry has shown robust growth over the past decade. As a result, the rationale for government intervention to provide funding today seems slim, as we discuss in more detail in the following.

7.5.2   Implications for Government Officials

Government officials and policy advisors are naturally concerned about spurring innovation. Encouraging venture capital financing is an increasingly popular way to accomplish these ends: numerous efforts to spur such intermediaries have been launched in many nations in Asia, Europe, and the Americas. A comprehensive review of these programs, and the lessons that can be learned from them, is beyond the scope of this chapter.

Nonetheless, five lessons can be highlighted:[15]

- Governments around the world today are seeking to promote entrepreneurial and venture capital activity, employing a variety of "stage setting" and direct strategies.
- These steps are supported by the historical record and theoretical arguments regarding the importance of such interventions in the development of entrepreneurial regions and industries.

15. These lessons are drawn from Lerner (2009).

- But these efforts are challenging. Governments cannot dictate how a venture market will evolve, and top-down efforts are likely to be unsuccessful.
- The same common flaws doom far too many programs. These reflect both poor design—reflecting a lack of understanding of the entrepreneurial process—and problematic implementation.
- Government must play a careful balancing act, combining an understanding of the necessity of playing a catalytic role with an awareness of the limits of its ability to stimulate the entrepreneurial sector.

As we have highlighted, venture capital is an intensely cyclic industry, and the impact of venture capital on innovation is likely to differ in this cycle. Yet government programs have frequently been concentrated during the time periods when venture capital funds have been most active and often have targeted the very same sectors that are being aggressively funded by venture investors.

This type of behavior reflects the manner in which such policy initiatives are frequently evaluated and rewarded. Far too often, the appearance of a successful program is far more important than actual success in spurring innovation. For instance, many "public venture capital" programs, such as the Small Business Innovation Research (SBIR) and the Advanced Technology Program (ATP) initiatives, prepare glossy brochures full of "success stories" about particular firms. The prospect of such recognition may lead a program manager to decide to fund a firm in "hot" industry whose prospects of success may be brighter, even if the sector is already well funded by venture investors (and the impact of additional funding on innovation quite modest). To cite one example, the ATP launched major efforts to fund genomics and Internet tools companies during periods when venture funding was flooding into these sectors (Gompers and Lerner 1999).

By way of contrast, the Central Intelligence Agency's In-Q-Tel fund appears to have done a much better job of seeking to address gaps in traditional venture financing (Business Executives for National Security 2001). The SBIR program provides another contrasting example. Decisions as to whether finance firms are made not by centralized bodies but, rather, devolved in many agencies to program managers who are seeking to address very specific technical needs (e.g., an Air Force research administrator who is seeking to encourage the development of new composites). As a result, many off-beat technologies that are not of interest to traditional venture investors have been funded through this program.

A far more successful approach would be to address the gaps in the venture financing process. As noted in the preceding, venture investments tend to be very focused into a few areas of technology that are perceived to have great potential. Increases in venture fundraising—which are driven by factors such as shifts in capital gains tax rates—appear more likely to lead to more intense competition for transactions within an existing set of technologies than to greater diversity in the types of companies funded. Policymakers may wish to respond to these industries' conditions by (a) focusing on technologies that are not currently popular among venture investors and (b) providing follow-on capital to firms already funded by venture capitalists during periods when venture inflows are falling.

More generally, the greatest assistance to venture capital may be provided by government programs that seek to enhance the demand for these funds, rather than the supply of capital. Examples would include efforts to facilitate the commercialization of early-stage technology, such as the Bayh-Dole Act of 1980 and the Federal Technology Transfer Act of 1986, both of which eased entrepreneurs' ability to access early-stage research. Similarly, efforts to make entrepreneurship more attractive through tax policy (e.g., by lowering tax rates on capital gains relative to those on ordinary income) may have a substantial impact on the amount of venture capital provided and the returns that these investments may yield. These less-direct measures may have the greatest success in ensuring that the venture industry will survive the recent upheavals.

In short, while government programs aimed at spurring venture capital and entrepreneurial innovation in alternative energy strive to produce a positive social rate of return, there are many challenges. The most effective programs and policies seem to be those which lay the foundations for effective private investment. Our analysis suggests that the market for venture capital may be subject to substantial "imperfections" and that these imperfections may substantially lower the total social gain achieved by venture finance. Given the extraordinary rate of growth (and now retrenchment) experienced by venture capital over the past decade, the most effective policies are likely those that focus on increasing the efficiency of private markets over the long term, rather than providing a short-term funding boost over a limited period.

References

Amis, David, and Howard Stevenson. 2001. *Winning angels.* New York: Pearson Education.
Barry, David G., and David M. Toll. 1999. Brentwood, IVP find health care, high tech don't mix. *Private Equity Analyst* 9 (1): 29–32.
Bingham, Kate, Nick Ferguson, and Josh Lerner. 1996. Schroder ventures: Launch of the Euro fund. Harvard Business School Case no. 9-297-026. Boston: Harvard Business School.
Business Executives for National Security. 2001. *Accelerating the acquisition and implementation of new technologies for intelligence: The report of the independent*

panel on the CIA In-Q-Tel venture. Washington, DC: Business Executives for National Security.

Cain, Walter M. 1997. LBO partnership valuations matter: A presentation to the LBO partnership valuation meeting. General Motors Investment Management Company. Mimeograph.

Devenow, Andrea, and Ivo Welch. 1996. Rational herding in financial economics. *European Economic Review* 40 (3–5): 603–15.

Estrin, Judy. 2008. *Closing the innovation gap: Reigniting the spark of creativity in a global economy.* New York: McGraw-Hill.

Fenn, George W., Nellie Liang, and Stephen Prowse. 1996. *The economics of the private equity market.* Washington, DC: Federal Reserve Board.

Foster management moves to dissolve consolidation fund. 1998. *Private Equity Analyst* 8 (December): 6.

Gompers, Paul, and Josh Lerner. 1997. Risk and reward in private equity investments: The challenge of performance assessment. *Journal of Private Equity* 1 (2): 5–12.

———. 1998. What drives venture capital fundraising? *Brookings Papers on Economic Activity, Microeconomics:* 49–192.

———. 1999. *Capital market imperfections in venture markets: A report to the Advanced Technology Program.* Washington, DC: Advanced Technology Program, U.S. Department of Commerce.

———. 2001. *The money of invention: How venture capital creates new wealth.* Boston: Harvard Business School Press.

Gorman, Michael, and William A. Sahlman. 1989. What do venture capitalists do? *Journal of Business Venturing* 4 (4): 231–48.

Gurung, Anuradha, and Josh Lerner. 2008. *The globalization of alternative investments.* Vol. 1. New York: World Economic Forum.

———. 2009. *The globalization of alternative investments.* Vol. 2. New York: World Economic Forum.

Hellmann, Thomas, and Manju Puri. 2000. The interaction between product market and financing strategy: The role of venture capital. *Review of Financial Studies* 13 (4): 959–84.

Kaplan, Steven N., and Per Strömberg. 2004. Characteristics, contracts, and actions: Evidence from venture capitalist analyses. *Journal of Finance* 59 (5): 2177–2210.

Kortum, Samuel, and Josh Lerner. 2000. Assessing the contribution of venture capital to innovation. *Rand Journal of Economics* 31 (4): 674–92.

Kreutzer, Laura. 2001. Many LPs expect to commit less to private equity. *Private Equity Analyst* 11 (1): 85–86.

Lerner, Josh. 1997. An empirical exploration of a technology race. *Rand Journal of Economics* 28 (2): 228–47.

———. 2009. *Boulevard of broken dreams: Why public efforts to boost entrepreneurship and venture capital have failed—and what to do about it.* Princeton, NJ: Princeton University Press.

Mollica, Marcos A., and Luigi Zingales. 2007. The impact of venture capital on innovation and the creation of new businesses. University of Chicago. Unpublished Manuscript.

New Energy Finance. 2008. *European and North American clean energy venture returns analysis.* London: New Energy Finance.

Poterba, James M. 1989. Venture capital and capital gains taxation. In *Tax policy and the economy.* Vol. 3, ed. Lawrence Summers, 47–68. Cambridge, MA: MIT Press.

Reyes, Jesse E. 1990. Industry struggling to forge tools for measuring risk. *Venture Capital Journal* 30 (September): 23–27.

Sahlman, William A., and Howard Stevenson. 1985. Capital Market Myopia." *Journal of Business Venturing* 1 (1): 7–30.

Summit's Jacquet departing to form own LBO firm. 1998. *Private Equity Analyst* 8 (May): 3–4.

Tyebjee, Tyzoon T., and Albert V. Bruno. 1984. A model of venture capitalist investment activity. *Management Science* 30 (9): 1051–66.

U.S. Department of Energy. 2008. *Renewable energy databook.* Washington, DC: GPO.

Wells, William A. 1974. Venture capital decision making. PhD diss., Carnegie-Mellon University.

World Economic Forum. 2009. Green investing towards a clean energy infrastructure. Project presented at the World Economic Forum Annual Meeting, Davos, Switzerland.

Contributors

Ashish Arora
Fuqua School of Business
Duke University
Box 90120
Durham, NC 27708-0120

Iain M. Cockburn
School of Management
Boston University
595 Commonwealth Avenue
Boston, MA 02215

Alfonso Gambardella
Department of Management and
    KITeS
Bocconi University
Via Roentgen 1
20136 Milan, Italy

Shane Greenstein
Kellogg School of Management
Northwestern University
2001 Sheridan Road
Evanston, IL 60208-2013

Rebecca M. Henderson
Harvard Business School
Morgan 445
Soldiers Field
Boston, MA 02163

Josh Lerner
Harvard Business School
Rock Center 214
Boston, MA 02163

David C. Mowery
Walter A. Haas School of Business
Mail Code 1900
University of California, Berkeley
Berkeley, CA 94720-1900

Richard G. Newell
U.S. Energy Information
    Administration, EI-1
1000 Independence Avenue, SW
Washington, DC 20585

Tiffany Shih
Department of Agricultural and
    Resource Economics
313 Giannini Hall
University of California, Berkeley
Berkeley, CA 94720

Scott Stern
MIT Sloan School of Management
100 Main Street, E62-476
Cambridge, MA 02142

Brian Wright
Department of Agricultural and
    Resource Economics
207 Giannini Hall #3310
University of California, Berkeley
Berkeley, CA 94720-3310

Jack Zausner
McKinsey & Company
5 Houston Center
Northwestern University
1401 McKinney Street
Houston, TX 77010

Author Index

Subject Index

Note: Page numbers followed by f *or* t *refer to figures or tables, respectively.*